Web Cartography

Web Cartography

developments and prospects

Edited by Menno-Jan Kraak
and Allan Brown

ITC
Division of Geoinformatics,
Cartography and Visualisation,
Enschede, The Netherlands

London and New York

First published 2001
by Taylor & Francis
11 New Fetter Lane, London EC4P 4EE

Simultaneously published in the USA and Canada
by Taylor & Francis Inc,
29 West 35th Street, New York, NY 10001

Taylor & Francis is an imprint of the Taylor & Francis Group

Publisher's Note
This book has been prepared from camera-ready copy provided by the
authors.

British Library Cataloguing in Publication Data
A catalogue record for this book is available from the British Library

Library of Congress Cataloging in Publication Data
Web cartography : developments and prospects / edited by Menno-Jan Kraak
and Allan Brown.
     p. cm.
     Includes index
1.Cartography – Data processing, 2.World Wide Web (Information
retrieval system)
I.Kraak, M. J. II.Brown, Allan.
                                    2001
GA102.4.E4 W43 2000   00-033763
526'.0285'4678 – dc21

Contents

Preface

The oldest maps were probably drawn in sand thousands of years ago and the most recent map is being created via the World Wide Web and sent to someone's mobile phone even while you are reading this page. Both have in common that they are short lived. Sand maps are real but are web maps real or just virtual? Obviously sand has many disadvantages and over the centuries many other more permanent and portable carriers have been used, among them clay tablets and paper. In the recent past map production was revolutionised by the advent of sophisticated computer software in the graphics industry and now an extremely rapid technological development is going on in the electronic dissemination of all kinds of information, including maps. At the present stage of this development for nearly all users, paper output has to be generated if an easily portable map is wanted. So rapid are the advances, however, that very soon after this book has been published we can expect wide access to portable, temporary maps on the new generation of Personal Digital Assistants coupled to mobile phone technology. The virtual, temporary characteristics of web maps can actually be seen as an advantage. They can be distributed easily and if changes occur, the supplier has to update only one single map and all the users immediately have access to the updated version. One other advantage is that most web maps are produced on demand. In some cases the user can decide which map to see and what its contents will be. Of course giving the user more control could be a disadvantage. What is the quality of the maps that might be produced?

In the book current developments on the Web relevant for those who deal with geospatial data are combined with recent trends in the world of geospatial data handling in general (e.g. national and global geospatial data infrastructures) and those in cartography in particular (interactive and dynamic mapping). This approach is presented as much as possible from the user point of view, since a clear shift from supply-driven cartography to demand-driven cartography is visible everywhere. Based on a classification of web maps, which includes a division into static and dynamic maps, the implications of the Web as a new medium to present cartographic products are discussed. Well-established theory and practice still form the basis of a successful map design, but special attention is given throughout the book to the new options the Web offers as well as to its limitations. Many of the points made in the more theoretical chapters are illustrated by examples from several time-sensitive applications that currently benefit from use of the Web. Any technical terms that might not be familiar to the general reader are fully explained.

The book itself takes advantage of the Web. It is accompanied by a website (http://kartoweb.itc.nl/webcartography/webbook). Many illustrations were produced specially for the book but in print they are of course static and view only. The website holds coloured versions of these illustrations and where relevant they are dynamic and interactive in order to demonstrate functionality discussed in the book. All hyperlinks (URLs) given in the book are accessible via this website and, even more important, are maintained. Here an analogy with maps in the sand is obvious. URLs disappear or change as quickly as they appear. The website also gives access to an online bibliography, more extensive and more up-to-date than the references found in the printed book. The website as such is a valuable extension to the book, but is of even more value if the book is used in an educational context. The lecture materials are always available anywhere and the map examples are interactive and dynamic, maybe even entertaining.

Our institute takes most of its students from the developing nations and we know from them and from our own travels that although access to the Web is at present very unevenly distributed, it is also quickly penetrating even the least developed areas of our planet. Via this book and its website we expect in particular to contribute to the mission of the institute; knowledge transfer and the development of geoinformation production, dissemination and use in developing countries. More generally, we hope that readers in any country with an interest in using and producing web maps will find something in the book to interest and stimulate them.

Acknowledgements

The Web Cartography project is an initiative of the Division of Geoinformatics, Cartography and Visualisation of the International Institute for Aerospace Survey and Earth Sciences (ITC), Enschede, The Netherlands, in the year 2000 celebrating its 50th anniversary. Nearly all staff members of the Division have been involved in writing chapters, preparing and drawing illustrations, collating and maintaining data for the URLs and the bibliography as well as in designing and maintaining the website. Our thanks as editors go to our colleagues for their enthusiastic cooperation, including to those not directly involved as chapter authors, namely Richard Knippers, Willy Kock and Ton Mank. Special thanks are also due to Wim Feringa for ensuring that all the illustrations were produced to the highest possible quality.

Menno-Jan Kraak
Allan Brown

Enschede, July 2000

CHAPTER ONE

Settings and needs for web cartography

Menno-Jan Kraak

1.1 NEW MAPPING ENVIRONMENTS

The World Wide Web (WWW) is the most recent new medium to present and disseminate geospatial data. Only a few players in the geosciences have not (yet) found a use for the WWW. In this process the map plays a key role, and has multiple functions. Maps can play the traditional role of providing insight into geospatial patterns and relations. Under these circumstances maps are used as they would in for instance an atlas or newspaper to present the structure of a city or the location of the latest earthquakes. However, because of the nature of the WWW the map can also function as an interface or index to additional information. Geographic locations on the map can be linked to for instance photographs, text, sound or other maps. The WWW is multimedia. Maps can also be used to preview geospatial data products to be acquired, when it comes to disseminating the data. This allows users to get a feel of the contents and coverage of a particular data set.

The purpose of this book is to provide information on the new opportunities and challenges offered by the WWW for cartography and related geosciences. It will describe the developments, changes and prospects of the mapping discipline in the framework of the WWW.

Cartography is about the design, production and use of maps. Web cartography is no different, but is restricted to the WWW as medium. As will become clear while reading this book most cartographic knowledge available before the advent of the WWW is still valid. However, as will be explained in later sections, although the WWW has some interesting new features it also has its limitations.

Maps can be defined as graphic representations of our environment. Definitions of maps have changed over time (see *URL 1.1*) due to changing insights, and web maps might add another one. However, web maps are maps, as described above, but they are presented in a web browser. The browser and the fact that most of these maps have to travel over networks put some constraints on the design and physical nature of web maps from the perspective of both the producer and the user. However, as Section 1.3 will show web maps do offer new opportunities.

The Web also puts new life into the map as a metaphor. The map is not only used in its natural environment to show locations, to explain geospatial patterns, or to navigate geographic space. New roles emerge. Maps will play a role in the geospatial data infrastructure needed to search for and disseminate geospatial data (Couclelis, 1998). Maps are now also used to navigate and map cyberspace and to map websites. Navigating the WWW is not easy, and after following a few links one might easily get lost, even though sites seem never further apart then 20 clicks

(Reka *et al.*, 1999). These new realms will result in some original and innovative map designs. Section 2.3 will elaborate on these traditional map functions in these new realms in more detail.

For those familiar with the WWW it will not be a surprise that web maps definitely will integrate cartography and multimedia. The use of maps on the Web implicitly hyperlinks these visualisations with other multimedia elements. The hypermap concept as introduced by Laurini and Milleret-Raffort (1990) and for instance elaborated for the Web by Kraak and Van Driel (1997), has become a common use of maps on the Web. From a cartographic perspective the WWW can be seen as a big hypermap. Its functionality is in line with the latest trends in cartography and related geodisciplines. This will be elaborated in Chapters 2 and 7, which focus on the keyword exploration.

Why is the WWW an interesting medium to present and disseminate geospatial data? The answer is that information on the Web is virtually platform-independent, unrivalled in its capacity to reach many users at minimal costs and easy to update frequently. Furthermore and more particularly in relation to maps, it allows for a dynamic and interactive dissemination of geospatial data, offering new mapping techniques and use possibilities not seen before with traditional printed maps, such as multimedia integration. Compared to on-screen maps it is the distribution factor that opens up new opportunities. These are all arguments that should appeal to geospatial data providers although some copyright issues and financial implications remain. With an increase in e-commerce (*URL 1.2*) the financial disadvantages might disappear. Copyright is a more awkward problem, since anything on the WWW is seen as public domain. Techniques such as watermarking (*URL 1.3*) the maps or securing the site through access restrictions (Finnish site *URL 1.4*) can help solve questions of copyright and use costs.

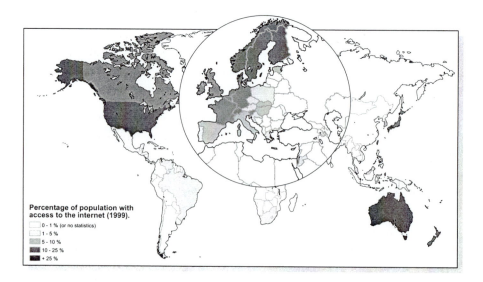

Figure 1.1 Web access (*URL 1.8*).

Technology and the user's location on Earth currently limit the interactivity and ease of use of the WWW. However, developments are proceeding quickly, and what is not possible today certainly will be tomorrow. Examples of advanced cartographic WWW options are animations to depict movement and change (*URL 1.5*), multimedia, i.e. maps combined with other graphics, sound, text and moving images (*URL 1.6*). Other examples are virtual environments that offer an interactive (realistic) 3D view of the landscape, for instance via flyby's and exploratory environments where geoscientists can work to solve their problems and make new discoveries (*URL 1.7*).

One has to keep in mind that despite its growth the WWW is accessible to a relatively limited group of people (*URL 1.8*), certainly not as yet evenly spread throughout the world as can be seen in Figure 1.1. The WWW is a fast medium used by impatient people. If the information takes too long to download, users will lose interest and go to other sites. The same people also expect up-to-date information. If the site appears older then a month it will be considered of no value. Most successful sites with a cartographic content are those offering time-sensitive information, such as the weather (see Chapter 11) or traffic (see Chapter 12) or those that allow the user to compose interactively location or route maps. These are points to consider by geospatial data providers when designing their site.

1.2 TYPES OF WEB MAPS

How can maps be put on the WWW? It is possible to distinguish between several methods that differ in terms of necessary technical skills from both the user's and the provider's perspective. The overview given in Figure 1.2 can only be a snapshot in time, since development on the WWW is tremendously fast. This "classification" is certainly not carved in stone, and one might easily find examples that would not fit these categories, or define new categories or combinations. Its objective is to give an overview of current possibilities, based on how the map image is used. The classification scheme will play a prominent role throughout the book, since all chapters are built on this scheme.

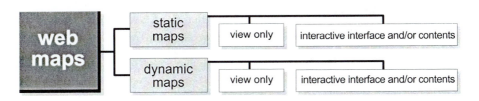

Figure 1.2 Classification of web maps (*URL 1.9*).

The scheme distinguishes between static and dynamic web maps. Each of these categories is further subdivided into view only and interactive maps. The most common map found on the WWW is the static view only map. Often the sources for these web maps are original cartographic products, which are scanned and put as bitmaps on the WWW. This form of presentation can be very useful, for instance, to make historical maps more widely accessible, as can be seen at a site where rare maps by the famous Dutch cartographer Blaeu can be viewed (*URL 1.10*). Often maps like these can only be seen and studied in the archives of libraries and map collections. The WWW offers these institutions the opportunity to share their cartographic heritage. Similarly, National Mapping Organisations (NMOs) could for instance offer their old map series. The static view only maps can also serve to give the visitors an impression of available products, and the NMOs could show details of their map series (see also Chapter 8). A problem with many of the static view only maps is that they have not been designed for the WWW. They have been scanned from atlases or other paper maps. Because of this their information density will be high and the maps may not be readable.

Static maps can also be interactive. These are the so-called "clickable" maps. The map can function as an interface to other data. Clicking on a geographic object could lead to other information sources on the Web (*URL 1.11*). These could be other maps, images or other web pages. Interactivity could also mean the user has the option to zoom and pan. Alternatively it could allow the user to define the contents of the web map by switching layers off or on (*URL 1.12*). Sometimes it is also possible to choose symbology and colours. NMOs could use their traditional index maps to offer information on individual map sheets (*URL 1.13*).

The WWW has several options to display dynamic processes via animations. The so-called animated GIF can be seen as the view only version of the dynamic maps. Several bitmaps, each representing a frame of the animation, are positioned after each other and the WWW browser will continuously repeat the animation. Famous (or notorious) examples are the commonly used spinning globes. Other applications are for example maps to depict the changing weather over the previous day (*URL 1.14*). Slightly more interactive versions of this type of maps are those to be displayed by media players, in AVI, MPEG or Quicktime format. Plugins to the WWW browser define the interaction options, which are often limited to simple pause, backward and forward. These animations do not use any environment parameters specific to the WWW and have equal functionality in the desktop environment. Interactive dynamics can be created by Java, JavaScript or via virtual environments in VRML or QuicktimeVR. VRML especially, that allows for the use of three-dimensional data, offers opportunities to define the travel path, and to make decisions on directions and height. A nice example of this is the VRML model of Schiphol Airport (*URL 1.15*). Furthermore, it offers the incorporation of links and thus becomes a more interactive "clickable animation".

1.3 WHAT IS SPECIAL ABOUT WEB MAPS?

Web maps definitely have some characteristics that make them different from paper maps or other on-screen maps. The scheme in Figure 1.3 structures these differences and also puts them into the perspective of this book. It is possible to look at the differences from a user point of view, or from that of the provider, while both perspectives have to consider the viewing environment and the map contents.

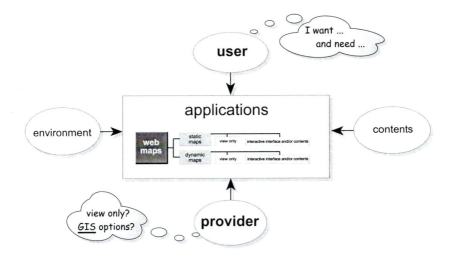

Figure 1.3 **What makes web maps special?**

The web map, in a specific application-dependent context, is influenced by four main actors. These are the user, the provider, the viewing environment and the map contents. It is the interaction among those actors that defines the typical web map appearance. The configuration outlined in the figure could also be applied to non-web maps. However, it is the WWW as medium that creates the specific environment with advantages and disadvantages for mapping.

If we look at the actors in the scheme from the web map perspective the unique nature of these maps can be made visible. The environment dictates that the web map should not be too large in both image and file size. Otherwise it is likely the user will be unwilling to wait for the map to download. The typical user surfing the Web has a relatively short span of attention. This put constraints on the contents as well. The graphic and information density should be low. This may sound like a disadvantage, but it guarantees a challenge to produce well-designed maps. Since the environment is very suitable for interaction it is possible to put all kinds of additional information behind the map image. This extra information could be made accessible via techniques such as mouse-over. When the user moves the cursor over the map and it hits a town symbol it could show the name of the town. This would obviate the need for the map to display many text symbols (see Chapter 7). One could also access a database by clicking objects in the map. The database could contain attribute data, but also multimedia elements such as sound or images. The need for these techniques will depend on the objectives of the providers. Do they just want to put information on the WWW in map form or do they offer geospatial data for use in for instance a GIS (Geographical Information System) environment? An example of the first could be a tourist organisation promoting several aspects of a particular region (*URL 1.16*). An example of the second could be a NMO offering full or sample data from its 1:50.000 database (*URL 1.17*).

All these options for interaction make it rather difficult for providers to know how their maps finally will appear, and who is looking at them. A different environment or a particular browser on a particular computer might result in a (slightly) different map than at the provider's own site. A solution could be just to provide view only maps, but in most cases that would not justify the provider's objectives. The provider has to decide what seems the best method to present the data. Questions such as which type of web map (static/dynamic) is applicable have to be answered. When it comes to production, the process will often have to be adapted compared to traditional mapping. An example could be if the provider intends to offer time-sensitive data.

1.4 THIS BOOK AND ITS WEBSITE

The book is split into four sections: the basics (chapters 1 & 2), using web maps (chapters 3 & 4), creating web maps (chapters 5, 6 & 7) and applications (chapters 8, 9, 10, 11 & 12). The first section provides the setting of web cartography and refreshes minds on what modern cartography is about, while focusing on the cartographic visualisation process. Figure 1.2 is the common thread throughout the book and it is revisited from the perspective of every chapter. The section on use comes before that on creation deliberately in order to avoid readers being influenced by a supply-driven view. Basically the book is written from a demand-driven perspective: I want…. therefore I need… One difficulty with this approach is that the users have to know something about what is possible before they can fully state their needs and it is unlikely that users are aware of the latest innovative technology available for web cartography. The section on using web maps pays special attention to their role in geospatial data infrastructures and examines different user profiles. Creating web maps includes chapters on basic cartographic design principles. This section also provides an overview of some technological opportunities and limitations and closes with specific design considerations for web maps. The application section deals with time-sensitive topics such as traffic, weather and tourism. Additionally NMOs and atlases receive special attention. The outlook section tries to look into the future and to describe a few trends likely to hit the Web soon. Among the topics treated are collaborative work and data mining, and the WWW and OpenGIS. The appendices go into some of the technical details required in a web cartography environment.

A book like this is not complete without an accompanying website. The site (http://kartoweb.itc.nl/webcartography/webbook) contains all illustrations found in this book in full colour. If the illustrations in the book refer to interactivity or dynamics these are available on the site version of the illustration. All links found in the book will be maintained at the site, new links will be added where appropriate and a searchable web cartography bibliography will be kept up-to-date. All the links were checked in the summer of 2000, but of course there can be no guarantee how long these links will remain accessible. If any URLs become inaccessible, and this will be checked regularly, an attempt will be made to put alternatives on the website.

URLs

URL 1.1   Map definitions <http://www.usm.maine.edu/~maps/essays/andrews.htm>
URL 1.2   E-commerce <http://www.nua.ie/surveys/index.cgi?f=FS&cat_id=14>
URL 1.3   Watermarking <http://www.maps.com/cgi-bin/magellan/Maps>
URL 1.4   Access restrictions <http://www.kartta.nls.fi/puzzle/>
URL 1.5   Animations
        <http://kartoweb.itc.nl/webcartography/webmaps/dynamic/di-example2.htm>
URL 1.6   Multimedia
        <http://kartoweb.itc.nl/webcartography/webmaps/dynamic/di-example3.htm >
URL 1.7   Flyby
        <http://kartoweb.itc.nl/webcartography/webmap/ dynamic/di-example3.htm >
URL 1.8   How many online <http://www.nua.ie/surveys/how_many_online/index.html>
URL 1.9   Classification of web maps
        <http://kartoweb.itc.nl/webcartography/webmaps/classification.htm>
URL 1.10 Blaeu's maps <http://odur.let.rug.nl/~welling/maps/blaeu.html>
URL 1.11 Clickable maps <http://www.britannica.com/bcom/eb/article/single_image
        /0,5716,367+bin%5Fid,00.html>
URL 1.12 Pan/zoom - switch layers <http://tiger.census.gov/cgi-bin/mapsurfer>
URL 1.13 Index <http://www.tdn.nl/anaindex.htm>
URL 1.14 Radar images weather <http://weerkamer.nl/radar/>
URL 1.15 Schiphol 3D-VRML
        <http://www.schiphol.nl/engine/index_def.html?lang=en&page_nr=590>
URL 1.16 Canterbury tour <http://www.hillside.co.uk/tour/map.html>
URL 1.17 Providing data <http://www.tdn.nl/demo_dl.htm>

REFERENCES

Couclelis, H., 1998, Worlds of information: the geographic metaphor in the visualisation of complex information. *Cartography and Geographic Information Systems*, **25** (4), pp. 209-220.

Kraak, M. J. and Driel, R. v., 1997, Principles of hypermaps. *Computers & Geosciences*, **23** (4), pp. 457-464.

Laurini, R. and Milleret-Raffort, F., 1990, Principles of geomatic hypermaps. In *4th International Symposium on Spatial Data Handling*, Zürich, pp. 642-655.

Reka, A., Hawoong, J. and Barabasi, A. L., 1999, Diameter of the World Wide Web. *Nature*, (401), pp. 130-131.

CHAPTER TWO

Trends in cartography

Menno-Jan Kraak

2.1 VISUALISATION PROCESS: CARTOGRAPHIC PRINCIPLES

Web cartography itself is of course a trend in cartography. However there are other recent trends that affect cartography and the way web cartography is developing. These have to do with the impact of visualisation and the need for interactivity and dynamics as well as the widespread use of GIS resulting in many more maps being produced by many more people. These people do this using the geospatial data infrastructure, an electronic highway for the use and dissemination of geoinformation. In the context of geospatial data handling, the cartographic visualisation process is considered to be the translation or conversion of geospatial data from a database into map-like products. Geospatial data handling stands for the acquisition, storage, manipulation and visualisation of geospatial data in the context of particular applications. During the visualisation process, cartographic methods and techniques are applied. These can be considered as a kind of grammar that allows for the optimal design, production and use of maps, depending on the application (see Figure 2.1). The following chapters will explain why and how this grammar has to be adapted and expanded in order to apply to the web map environment.

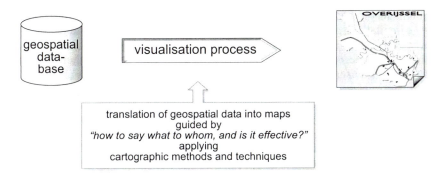

Figure 2.1 The cartographic visualisation process.

The producer of maps may be a professional cartographer, but may also be a discipline expert making maps of for instance, forestry stands using remote sensing images or mapping a city's employment statistics. To be able to execute the translation from geospatial data into graphics, it is assumed here that data are available and that the geospatial database is well structured. This should ensure that the data can respond to all the queries required by the application field for by a whole set of cartographic tools and theory as described in cartographic textbooks (Robinson *et al*, 1995; Kraak and Ormeling, 1996)

During recent decades, many others have become involved in making maps. The widespread use of GIS has increased the number of maps created tremendously (Longley *et al*, 1999). The WWW has an even greater impact on the number of maps created (see Chapter 4). Even the spreadsheets used by most office workers today have mapping capabilities, although most of them are probably not aware of this (Whitener and Creath, 1997). Many of these maps are not produced as final products, but rather as intermediate products to support the user in his or her work dealing with geospatial data. The map, as such, has started to play a completely new role: it is not just a communication tool but also a tool to aid the user's (visual) thinking process.

This process is being accelerated by the opportunities offered by hardware and software developments. These have changed the scientific and societal needs for geo-referenced data and, as such, for maps. New media such as CD-ROMs and particular the WWW not only allow for dynamic presentation but also for user interaction (Cartwright *et al*., 1999). Users expect immediate and real-time access to the data; data that have become abundant in many sectors of the geoinformation world. This abundance of data, seen as a paradise by some sectors, is a major problem in other sectors. Currently, one lacks the tools for user-friendly queries and retrieval when studying the massive amount of data produced by sensors, and now available via the WWW. A new branch of science is currently evolving to solve this problem of abundance. In the geodisciplines, it is called visual geospatial data mining (see Chapter 13). The developments have given the word *visualisation* an enhanced meaning. According to the dictionary, it means "make visible" and it can be argued that, in the case of geospatial data, this has always been the business of cartographers. However, progress in other disciplines has linked the word to more specific ways in which modern computer technology can facilitate the process of "making visible" in real time. Specific software toolboxes have been developed, with functionality based on two key words: interaction and dynamics. A separate discipline, called scientific visualisation, has developed around it (McCormick *et al*, 1987). This is having a major impact on cartography as well (see MacEachren and Kraak, 1997). If applied in cartography it offers the user the possibility of instantaneously changing the appearance of the map. Interacting with the map will stimulate the user's thinking and will add a new function to the map. As well as communication, it will prompt thinking and decision-making. Developments in scientific visualisation stimulated DiBiase (1990) to define a model for map-based scientific visualisation. It covers both the communication and thinking functions of the map. Communication is described as "public visual communication" since it concerns maps aimed at a wide audience. Thinking is defined as "private visual thinking" because it is often an individual playing with the geospatial data to determine its significance (see Figure 2.2). See also Card *et al*. (1999).

It is obvious that presentation fits into the traditional realm of cartography, where the cartographer works on known geospatial data and creates communicative maps. These maps are often created for multiple uses. Exploration, however, often involves a discipline expert creating maps while dealing with unknown data. These maps are generally for a single purpose, expedient in the expert's attempt to solve a problem. While dealing with the data, the expert should be able to rely on cartographic expertise, provided by the software or some other means. The nature of this cartographic expertise will be discussed later

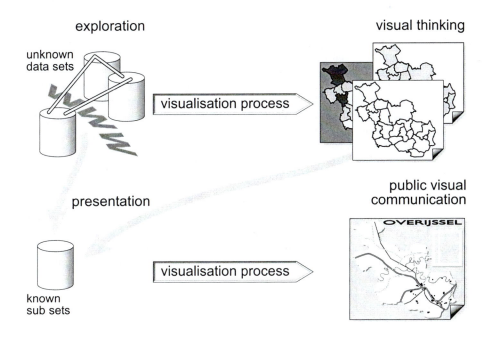

Figure 2.2 Visual thinking and visual communication.

Exploration in cartography is relatively new to the discipline. By understanding what happens, especially in the realm of visual thinking, at the exploratory phase of geospatial data handling cartographers can judge how well prepared the discipline is. Exploration means working with unknown data. However, what is unknown for one is not necessarily unknown to others. For instance, browsing in Microsoft's Encarta World Atlas is an exploration for most users because of its wealth of information. With products like these, such exploration takes place within boundaries set by the producers. Cartographic knowledge is incorporated in program wizards resulting in pre-designed maps. Some users feel this to be a constraint, but those same users will no longer feel constrained as soon as they follow the web links attached to this electronic atlas (see also Chapter 10). This example shows that the environment, the data and the type of users influence one's view of what exploration entails. However, for those with some web surfing experience, exploring might be equal to getting lost, or end up in the realm of the dead-end streets of links no longer available.

Getting lost or confused is not what the person who is trying to solve a particular geo-problem and who is exploring various geospatial databases is looking for. How can web cartography help? Let us revisit the phrase in Figure 2.1 that drives the visualisation process, *"How do I say what to whom, and is it effective",* and see what will be different considering exploration and the Web.

"*How*" still represents the cartographic methods and techniques. However, recent technological innovations offer challenges and opportunities. One can think of animation, the application of the third dimension and virtual reality, multimedia, etc. In addition, the user environment changes. WWW is a whole new environment for mapping. Maps will be the primary tools in an interactive, real-time and dynamic environment, used to explore geospatial databases that are hyperlinked together via the WWW. Questions that arise are "Which of the traditional tools are still valid", "How do the new tools work", and "What function can they play during the visualisation process?" "*I*" still is the mapmaker, but more often a geoscientist. "*What*" no longer represents a relatively well-defined and known data set; at least, certainly not from the user perspective. This is a good example that demonstrates the usefulness of meta data. "*Whom*" seems to be simpler than before. It is not a relatively well-defined user group, but the same person represented by "*I*", the expert geoscientist in the role of cartographer. "*Effective*" raises some interesting questions. If a map is considered effective, is it because of the efficient graphics or because of the geoscientist's clever thinking?

Although there are similarities between the nature of the visualisation process in an exploratory environment and in a presentation environment, there are also major differences. One of the hidden trends is clearly a shift from supply-driven cartography to a demand-driven approach. This urges the cartographer to concentrate on three major challenges for the future development of cartography (Kraak, 1998). They deal with:

- tool development
- geospatial data access
- effectiveness.

The tools to be developed, or to be improved, should allow the user to look at geospatial and other geo-referenced data in any combination, at any scale, with the aim of seeing or finding geospatial patterns (which may be hidden). One of the first concepts of visual geospatial data exploration was introduced by Monmonier (1989) when he described the term "brushing" (see Figure 2.3). This is when the selection of an object in a map automatically highlights the corresponding elements in the other graphics. Depending on the view in which one selects the object, there is geographical brushing (clicking in the map), attribute brushing (clicking in the diagram or table), and temporal brushing (clicking on the time line). As such, the user gets an overview of the relation among geographic objects based on location, characteristics and time. Most of today's geospatial data viewing software, such as ArcView, has elaborated on these principles. Recent developments are described in MacEachren and Kraak (1997) and in Kraak and MacEachren (1999). Cartographic Visualisation in this literature is also referred to as GVis or Geographic Visualisation. The tools have no purpose if the user does not have access to a geospatial database or, more likely, multiple databases. Under the best circumstances, users can access these within their own organisation. However, exploring geospatial data nearly always involves accessing different types of data from different sources. The key to data access will be the WWW.

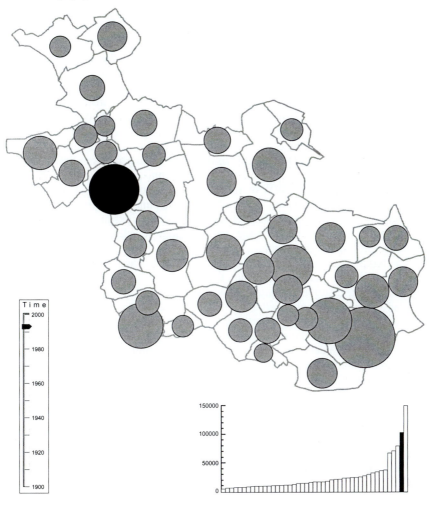

Figure 2.3 Visual geospatial data exploration.

2.3 NEW CARTOGRAPHIC REALMS: MAPPING CYBERSPACE

In the past cartography played an important role in the exploration of the world. Maps were used to chart unknown territories. A new phase in mapping the unknown has recently started. This does not refer to the cartographic or geographic exploration discussed in the previous paragraph. It deals with the mapping of cyberspace. Cyberspace is "a consensual hallucination experienced daily by billions of legitimate operators, in every nation, by children being taught mathematical concepts.... A graphical representation of data abstracted from the banks of every computer in the human system. Unthinkable complexity. Lines of light ranged in the non-space of the mind, clusters and constellations of data. Like city lights, receding..." as worded by William Gibson, who coined the term cyberspace in 1984 (Gibson, 1984). Currently the geography of cyberspace receives wide attention. Cyber geography is defined as "the study of the geospatial nature of computer communications networks, particularly the Internet, the World Wide Web and other electronic "places" that exist behind our computer screens, popularly referred to as *cyberspace*. Cyber geography encompasses a wide range of geographical phenomena from the study of the physical infrastructure, traffic flows, the demographics of the new cyberspace communities, to the perception and visualisation of these new digital spaces" (*URL 2.2*). Several books have been published in this field (Chen, 1999; Crang, Crang *et al.*, 1999; Crampton, 2000; Hillis, 2000; Kitchin and Dodge, 2000), as well as many papers (Batty and Barr, 1994; Batty, 1997; Jiang and Ormeling, 1997; Brunns, 1998; Dodge, 1998; Jiang and Ormeling, 1999). The WWW itself is a good source for references to and examples of cyber geography. A good starting point is Martin Dodge's cyber geography website (*URL 2.3*). Besides an atlas that provides an overview of visual views on many aspects of cyberspace, and a regular newsletter, this website offers convenient entry into cyberspace.

On looking at the images on these pages it is clear that many alternative mapping techniques have been used. Cyberspace is mapped preferably in 3D, using animation, different projection and many colours. An approach like this is simulated by Keller and Keller (1992), who distinguish three steps in the visualisation process: the first identifies the visualisation goal; the second removes mental roadblocks; and the third designs the display in detail. Here the first deals with visualising aspects of cyberspace. In the second step, Keller and Keller suggest taking some distance from the discipline in order to reduce the effects of traditional constraints. Why not choose an alternative mapping method? For instance, show an animation or VRML landscape. Because many non-cartographers are currently involved in mapping cyberspace finding alternative mapping methods is not strange. New, fresh, creative graphics are the result. They might also offer different insights and would probably have more impact than traditional mapping methods. During the third step, which is especially applicable in an exploratory environment, one has to decide between mapping data or visualising phenomena. However, as will be demonstrated in Chapter 5 "traditional" mapping rules still have their validity.

Although cyberspace is less tangible then the real world outside, geography can be used to increase insight. First it has to be defined what to map about cyberspace. Dodge's definition of cyber geography in the previous section has links to the components of geospatial data: location, attributes and time (*URL 2.4*). Considering location, the topics to map could be elements of the WWW infrastructure, its hosts/servers, their connections, and their users. An example of the last is a map showing the penetration of the Internet as shown in Figure 1.1. Locations can be mapped absolutely (for instance in longitude and latitude or locations of cables – *URLs 2.5 + 2.6*) or relatively (for instance topological relations – *URL 2.7*). Several tools exist to help one map the WWW (*URLs 2.8 + 2.9*). Specific tools are those that try to map the cyberspace routes followed while connecting to other sites (*URLs 2.10, 2.11 + 2.12*). Thinking of attributes one can map the characteristics of the infrastructure. This might include the intensity of use, the capacity etc. (*URL 2.13*). Other examples are the so-called Internet weather maps showing real-time Internet traffic (*URLs 2.14, 2.15 + URL 2.16*).

Mapping time is especially relevant for the WWW, since change is almost synonymous with the WWW. The time-related topic will vary depending on the scale one uses to look at time. It could be the time it takes to make connections to particular sites (use the WWW when North America is asleep) or the time it takes before there is WWW access in African countries or developments in the USA (*URLs 2.17 + 2.18*). Of course, most maps hold some kind of combination of location, attribute and time. Sometimes the maps represent a virtual world like Alpha World (*URL 2.19*). The online journal Mappa Mundi (*URL 2.20*) offers a flavour of all aspects of cyber space.

Browsing the cyber atlas does not reveal "new" map types. One finds map types like dot maps, choropleth maps, quantitative dot symbol maps and movement maps. However many maps are in 3D, in a landscape or prism format. Another observation is that data are mapped in schematic formats, just showing topology. There is a clear analogy with the well-known schematic London Underground map. This underground map and many of its derivatives function well because the user does not need a notion of the "where" in relation to the real world while travelling. Looking at the many examples dealing with mapping cyberspace and comparing them with the web map types found in the scheme in Figure 1.2 one finds that most of them represent the static view-only maps. Only a few allow for interaction or have a dynamic component. This is partly due to the fact that most of the maps are produced based on monitoring events on the WWW and are presented as final products. Within the realm of web cartography there are some interesting opportunities to add interaction and dynamics.

Maps are not only used to visualise cyberspace: they can also be useful in navigating cyberspace or even presenting a single website. Readers with some experience in surfing the WWW know that it is rather easy to get lost. According to Kahn (1995) the user has to be supported both on a global as well as on a local level. Local navigation implies following a link between two nodes, for instance by clicking the maps to move to a photograph of the area. Global navigation refers to movements spanning many nodes, for instance by navigation in an online atlas. Both levels of navigation have to be present. When the map metaphor is used for navigation schematic maps are often used.

URLs

URL 2.1 Map examples
          <http://kartoweb.itc.nl/webcartography/webmaps/classification.htm>
URL 2.2 Cyberspace <http://www.cybergeography.org/about.html>
URL 2.3 Cyber geography <http://www.cybergeography.org/>
URL 2.4 Cyber atlas <http://www.cybergeography.org/atlas/atlas.html>
URL 2.5 Location < http://www.cybergeography.org/atlas/kdd_asia_large.gif >
URL 2.6 Connections <http://www.telegeography.com/Publications/map97.html>
URL 2.7 In fractals <http://www.cs.bell-labs.com/who/ches/map/index.html>
URL 2.8 Tools to map <http://www.caida.org/tools/visualization/mapnet/>
URL 2.9 Find coordinates <http://www.ckdhr.com/dns-loc/>
URL 2.10 Neotrace <http://www.neotrace.com/>
URL 2.11 Visual trace <http://www.visualroute.com/>
URL 2.12 Gtrace < http://www.caida.org/tools/visualization/gtrace/ >

URL 2.13 Hosts <http://www.mids.org/mapsale/world/>
URL 2.14 Internet weather <http://www.mids.org/weather/>
URL 2.15 USA weather <http://hydra.uits.iu.edu/~abilene/traffic/abilene.html>
URL 2.16 European weather <http://sigma.dante.org.uk/mystere/mesh-map/>
URL 2.17 Africa and WWW <http://www3.sn.apc.org/africa/afrmain.htm>
URL 2.18 USA and WWW <http://skew2.kellogg.nwu.edu/~greenste/research.html>
URL 2.19 Alphaworld <http://www.activeworlds.com/satellite.html>
URL 2.20 Mappa Mundi <http://mappa.mundi.net/maps/>

REFERENCES

Batty, M., 1997, Virtual Geography. *Futures*, **29** (4/5), pp. 337-352.
Batty, M. and Barr, B., 1994, The electronic frontier: exploring and mapping cyberspace. *Futures*, **26** (7), pp. 699-712.
Brunns, S.D., 1998, The internet as "the new world" of and for geography: speed, structures, volumes, humility and civility. *GeoJournal*, **45** (1/2), pp. 5-15.
Card, S.K. MacKinlay, J.D and Shneiderman, B., 1999, *Readings in information visualization: using vision to think*, (San Francisco: Morgan Kaufmann publishers).
Cartwright, W., Peterson, M., Gartner, G., Eds., 1999, *Multimedia Cartography*, (Berlin: Springer)
Chen, C., 1999, *Information visualisation and virtual environments*, (Berlin: Springer-Verlag).
Crampton, J.W., 2000, *The geographies of cyberspace and places*, (Edinburgh: Edinburgh University Press).
Crang, M., Crang, P., May, J., 1999, *Virtual Geographies: bodies, space and relations*, (London: Routledge).
DiBiase, D., 1990, Visualization in earth sciences. *Earth & Mineral Sciences, Bulletin of the College of Earth and Mineral Sciences*, **59** (2), pp. 13-18.
Dodge, M., 1998, The geographies of cyberspace. *94th Annual meeting of the Association of American Geographers*, Boston, MA.
Gibson, W., 1984, *Neuromancer*, (NewYork: Ace Books).
Harley, J.B., 1991, Can there be a cartographic ethics? *Cartographic Perspectives*. (10), pp. 10-11.
Harpold, T., 1999, Dark continents: a critique of internet metageographies. *Post modern culture*, **9** (2).
Hillis, K., 2000, *Digital sensations: space, identity, and embodiment in virtual reality*, (Minneapolis: University of Minnesota).
Jiang, B. and Ormeling, F.J., 1997, Cybermap: the map for cyberspace. *Cartographic Journal*, **34** (2), pp. 111-116.
Jiang, B. and Ormeling, F.J., 1999, Mapping Cyberspace: Visualising, Exploring and Analysing Virtual Worlds. In *Proceedings of the 19th International Cartographic Conference ICC99, Ottawa*, (Ottawa: Canadian Institute of Geomatics), pp. 629-636.
Kahn, P., 1995, Visual cues for local and global coherence in the WWW. *Communications of the ACM*, **38** (8), pp. 67-69.
Keller, P.R. and Keller, M.M, 1992, *Visual cues, practical data visualization*, (Piscataway: IEEE Press).

Kitchin and Dodge, M., 2000, *Mapping cyberspace*, (London: Routledge).

Kraak, M.J., 1998, Exploratory cartography, maps as tools for discovery. *ITC Journal* (1), pp. 46-54.

Kraak, M.J. and MacEachren, A.M., 1999, Visualisation for exploration of geospatial data. *International Journal of Geographical Information Sciences*, **13** (4), pp. 285-287.

Kraak, M.J. and Ormeling, F.J., 1996, *Cartography, the Visualization of Geospatial Data*, (London: Addison Wesley Longman).

Longley, P., Goodchild, M., Maguire, D.M. and Rhind, D., Eds., 1999, *Geographical Information Systems: Principles, Techniques, Management, and Applications*, (New York: J. Wiley and Sons).

MacEachren, A.M. and Kraak, M.J., 1997, Exploratory cartographic visualization: advancing the agenda. *Computers & Geosciences*, **23** (4), pp. 335-344.

McCormick, B., DeFanti, T.A. and Brown, M.D., 1987, Visualisation in Scientific Computing. *Computer Graphics*, **21** (6).

Monmonier, M., 1989, Geographic Brushing: enhancing exploratory analysis of the scatterplot matrix. *Geographical analysis*, **21** (1), pp. 81-84.

Morrison, J.L., 1997, Topographic Mapping for the Twenty First Century. In *Framework of the World*, edited by Rhind, D., (Cambridge: Geoinformation International), pp. 14-27.

Robinson, A.H., Morrison, J.L. Muehrcke, P. C., Kimerling, A. J., Guptill, S. C., 1995, *Elements of Cartography*, (New York: J. Wiley and Sons).

Whitener, A. and Creath, B., 1997, *Mapping with Microsoft Office, using maps in everyday office operations*, (Santa Fe: OnWord Press).

Wood, D., 1992, *The power of maps*, (London: Routledge).

CHAPTER THREE

Use of maps on the Web

Corné P.J.M. van Elzakker

3.1 INTRODUCTION

In cartographic literature there is sometimes confusion about the terms use and user. And this confusion even grows as a consequence of modern developments in cartographic visualisation, as described in the preceding chapters of this book.

In some cases, authors refer to the producers who use maps to disseminate geospatial data, for instance through the WWW. In a similar way, website designers may use a clickable map as interface to the information residing on the site, be it geospatial or not (*URL 3.1*). Sometimes, maps are also applied as means of organisation of the wealth of information on the WWW as a whole, or part of it, in cases where the geospatial component does not exist (*URL 3.2*) or is not immediately obvious or of prime importance (*URL 3.3*). These kinds of web index or 'category' maps appear to function in one way or another in situations where 'surfers' are browsing the Web for no specific reason (Chen *et al.*, 1998). However, somebody interested in a piece of music composed by Strauss will normally not search and call up that information by looking for and clicking on a map with the geographical location of the server that actually carries it.

When reference is made to the user, cartographers normally have in mind the persons who are actually using maps (including, perhaps, index maps or maps as interfaces) to find answers to the essentially geographical questions they have. And this is also the perspective from which Chapters 3 and 4 of the book have been written. But, this being the case, web map use in a rather broad sense is meant, not limited to just the use of maps which are already displayed on the computer screen by a particular map user. In fact, being considered is the whole process of using the WWW and a web browser to retrieve (geographic) information which is (or can or could be) communicated by cartographic means. At first, the Web is used to find and retrieve data which may be an answer to the geographical questions users have. These geographic data may already come in the form of ready-made cartographic displays or they still have to, or could, be cartographically visualised by the users themselves. Thereafter, the users may actually use the maps thus generated to obtain the information required. Realising that WWW users are first of all looking for answers to their (geographical) questions, it should also be noted that users often do not search for a particular map, but the map display is offered as a (possible) answer to a more general question like: "Where can I find a Chinese restaurant?" The Dutch version of the Yellow Pages (*URL 3.4*) offers the users, next to address and other textual information and perhaps fully unexpectedly for the user, a map ("Toon Kaart" = show map) with the locations of all Chinese restaurants in the region specified; or a map showing the location of the restaurant selected; or even (if the business was willing to pay for it) a map (next to a textual description) showing the route to the restaurant selected from the place where the user is staying. In this case, the user did not ask for a map, but was only looking for a place to eat. And once the map is offered, the user will be pleased with this way of supplying an answer to the question, particularly also if cartographic functionality like panning and zooming is provided (as is the case on the Dutch Yellow Pages).

3.2 WHAT'S THE USE OF WEB MAPS?

In the first two chapters of this book it has already been demonstrated that maps are very effective tools for the transfer of geospatial data. Maps may provide insights and overviews that cannot be obtained with other means of communication. However, currently on the Web only limited use is made of the cartographic potential (Van Elzakker & Koussoulakou, 1997). Many of the subjects which are to be found on the WWW have a strong geospatial component, but are not always presented in the best way: common examples are tourism and travel sites (see Chapter 9), many of which contain only descriptive text and photographs. Although the environment of the Web is graphically, and thus cartographically, appealing, in this sense it often still is rather primitive.

Nevertheless, many maps already reside on the WWW or users have access to geospatial data which may be cartographically visualised during a web session. If you would ask somebody with experience in using web maps what the advantages of this new medium are, in comparison with other media for the dissemination and use of geospatial data, there is a fair chance that such a person would make a comparison with traditional paper maps and, perhaps, atlases. Advantages that could be mentioned are the possibilities to easily look up places on a map, to pan and to zoom, to select map layers to be displayed, to make use of hyperlink functionality and of integrated multimedia components (like sound, pictures and video or animations). What should be realised, however, is that these new possibilities for dynamic interaction with geospatial data and the generation of tailor-made cartographic displays were already there in computerised GIS environments and with the introduction of, first, diskettes and then CD-ROMs as carriers of geographic and cartographic information instead of the traditional paper. The cartographic functionality offered at Microsoft's website Expedia Maps (*URL 3.5*) is not very much different, for instance, from what was already possible with Microsoft's Encarta World Atlas on CD-ROM. In a similar way, the route planning functionalities offered at many websites nowadays (see Chapter 12) were first, and still are, available in route planners on CD-ROM.

What then are the real advantages of the new WWW medium from the perspective of the user of web maps and compared to, say, a medium like CD-ROM for the dissemination of geographic and cartographic information? The advantages of the WWW as a medium may be summarised under two main headings:

- Accessibility
- Actuality

   A user with access to the WWW has, in principle, access to an enormous wealth of information from his or her PC at home or at the workplace. Information, including web maps, is easily accessible through user-friendly web browsers, 24 hours a day and not hindered by political and geographical boundaries. Through (hyper)linking users also have limitless access to much more information than could ever be carried on a single CD-ROM; the Web may provide a quick answer to almost any geographical question. Users also do not have to leave their homes or workplaces to buy a CD-ROM in a shop with limited opening hours, nor do they have to worry about installing the CD-ROM on their PCs. In Section 1.2 the example was already given of making accessible to the users through the WWW scanned copies of rare historical maps that are only available in one or a few map libraries in the world with perhaps very restricted access because of their vulnerability (see *URL 1.10* or *URL 3.6*). The accessibility of the medium also creates possibilities for public participation and collaborative cartographic visualisation in, for instance, physical planning procedures (Krygier, 1999). Another aspect of the accessibility and a big advantage to the user is that a lot of the information on the WWW is still available free of charge (see Section 4.3), although it should, of course, be realised that suitable hardware, software and an Internet connection are required.

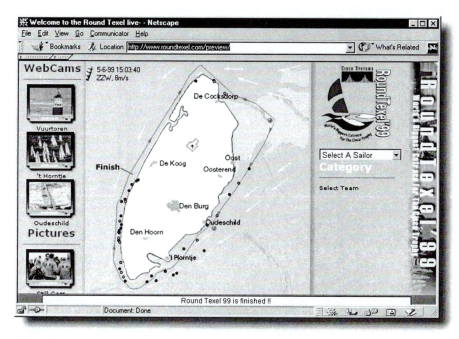

Figure 3.1 Real-time geographic information dissemination: the Round Texel catamaran sailing race. (Source: Cisco Systems and O.G. Haaglanden, studierichting Technische Informatica) (*URL 3.11*).

One of the most serious problems of traditional cartography was to keep maps up-to-date. Due to the lengthy production process, sometimes a paper map was only made available to the users years after the initial data collection. By that time some of these outdated maps were already of limited use. With the introduction of electronic mapping the production process could be speeded up somewhat, but the problem of actuality remained. A new edition of a route planner on CD-ROM will not be published every month, and even if it were, users would not be willing to buy a new version that often. The WWW, however, makes it possible to supply the users with really up-to-date geographic and cartographic information. Examples are sites with web maps showing the weather situation ("Are showers of rain approaching?" (*URL 3.7*), or, more seriously, "What is the track of that dangerous hurricane?" (*URL 3.8*). Also see Chapter 11). Other examples are sites with web maps showing (almost) real-time traffic information related to road construction work, incidents and traffic congestion (*URL 3.9*, see Chapter 12 for more examples). Ultimately, the limit to the speed of revision is the speed of the data transfer through the Internet. A step further is to make predictions of traffic conditions online available to the user. The University of Duisburg in Germany, for instance, has created a computer simulation model by which traffic flows are predicted for the next 10 to15 minutes, on the basis of measurements of current traffic intensities (*URL 3.10*). And a next step would be, of course, to incorporate this kind of information in the routeplanners that are available on the Web. A last example of the unprecedented potential of the WWW to provide really up-to-date information by means of web maps is formed by sites that keep users informed of recent developments in news and sports. A nice illustration are the websites that inform people at home about the actual position of boats participating in sailing races. The positions of the boats are recorded by means of GPS techniques and continuously plotted on sea charts that can be consulted on the Web (Figure 3.1). During the first ten days of the Route du Rhum 98 sailing race, more than 5 million maps were displayed online in this way (Baumann, 1999).

These are all examples of new possibilities of a new medium to supply (almost) real-time geographic information by means of web maps. But geographic information that is somewhat less dynamic (e.g. tourist maps (see Chapter 9) or topographic base maps (Chapter 8)) may now also be supplied to the user in more up-to-date form than ever before. Easily accessible and up-to-date topographic base map information is urgently required, for instance, for rescue workers in case of natural disasters. But also for less urgent situations, it is important, indeed, to constantly keep the web maps and the geographic information as up-to-date as possible, as it may be expected that the users will become more critical in this respect than they were ever before and that they will lose their confidence in websites that are not kept up-to-date regularly.

In this section, the use of maps on the Web has been compared with maps on other information carriers like paper and CD-ROM. In relation to the advantages of the WWW medium, it should finally be mentioned that the Web may also be used in combination with paper maps and atlases or, for instance, routeplanners or electronic atlases on CD-ROM, by allowing and providing updates in between the publication of new editions of these more static products. An example is the National Geographic Atlas of the World on paper in combination with the Map Machine website (*URL 3.12*).

3.3 MODES OF USING WEB MAPS

In Section 1.2 a classification of web maps was presented that forms one of the starting-points for the contents of this book (see Figure 1.2). The subdivision made at the lowest hierarchical level of this classification (view only versus interactive interface and/or contents) is a subdivision made from the perspective of the web map user.

Another way of looking at the modes of using web maps – following what has been stated about visualisation strategies in Section 2.2 – is to consider map use goals as positioned in the so-called "map use cube", originally conceived by MacEachren (1994) (see Figure 3.2).

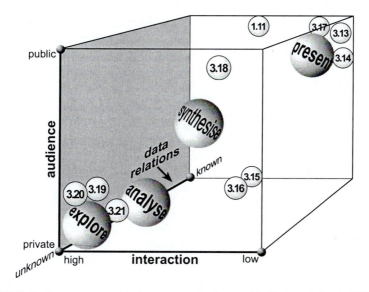

Figure 3.2 Goals of map use arrayed in the map use cube (source: MacEachren & Kraak, 1997). By means of illustration some of the examples (*URLs*) that are mentioned in the text have been positioned in the cube as well.

The three axes in the cube represent three different map use continua:
- from map use that is private (where an individual generates a map for his / her own needs) to public (where ready-made maps are made available to a group of users);
- from map use that is directed towards revealing unknowns to presenting knowns;
- from use with a high human-map interaction (where the user can manipulate maps substantially) to low interaction (where the user has limited ability to change the presentation).

Maps, including the maps generated in WWW sessions, may occupy any position in the three-dimensional space defined by these axes, depending upon what a user does with the maps for what purpose. MacEachren and Kraak (1997) recognised four map use goals; to explore, to analyse, to synthesise and to present and they positioned these in the cube (see Figure 3.2). However, in principle, web maps may also occupy other positions in the cube, depending on the typical use characteristics.

The static view only scans of existing paper maps which were put on the Web first several years ago occupy a position close to the "present" ball in the cube. Many of these maps can still be retrieved through the Perry Castañeda Library Map Collection site (*URL 3.13*). Typically, they were designed for a wide group of users and for a general purpose. Later on, the dynamic equivalents of these view only maps were introduced on the Web (*URL 3.14*). They may occupy the same position in the map use cube.

As the Web typically is a medium for private use, many cartographic sites can be found near the base of the cube. Through these sites maps may actually be created by an individual user to suit his or her private needs. When these possibilities for online map creation are limited to the selection of an area, a projection method, switching layers of map details on and off and the design of the symbols representing these details, including the selection of colours, we are dealing with medium interactivity and the presentation of known geographical data relations (*URL 3.15*). This implies a position near the middle of the bottom side at the back of the cube. In a sense, clickable maps or hypermaps for public use may also be regarded as a kind of medium interactive map, occupying a high position near the back of the cube (*URL 1.11*). In other interactive maps on the Web, users may change the area portrayed (panning) or the scale (zooming) (*URL 3.16*) and some user-friendly sites even allow the user to change the orientation of the map display (e.g. North or destination at the top of a route map - see Chapter 12).

The presenting knowns to revealing unknowns axis of the map use cube also reflects very well different conditions of map use through the Web. On the "presenting knowns" end, users know exactly what geographical information they want and often also what map on which website supplies that information to them. For example, the site of the Dutch High-Speed Line Project (*URL 3.17*) contains maps showing the planned routes of the railway line. These maps may also be positioned close to the "present" ball in the map use cube. On the front side of the cube we may find the Web surfers who may not exactly know what they are looking for and browse, for instance, through one of the atlases on the Web. For instance, the Lycos World Atlas (*URL 3.18*) may be positioned somewhere near the middle/right of the top of the front side of the cube.

Currently, in web cartography as in cartography in general, lots of interesting developments are taking place in the left hand bottom front corner of the map use cube (see Section 2.2). This is the position of exploratory cartography: map use in the private (revealing unknowns) and high human map interaction corner of the cube. Because of further developments in the client-server architecture (see Chapter 6), it becomes ever more possible for web map users to explore and really interact with certain geospatial datasets, while making use of modern cartographic visualisation techniques, in order to gain insight into these unknown datasets. In such cases, it may be possible to manipulate (e.g. classify) the data, choose different cartographic representation methods and visually compare the resulting map displays. As such, online visual exploration may be followed by downloading the geospatial data for analysis locally. In this respect, there are some interesting examples on the Web relating to the exploration of census data. The CIESIN

DDViewer can be used to explore cartographically and to calculate statistics for 220 demographic variables from the 1990 US Census (*URL 3.19*). In the United Kingdom, the Census Dissemination Unit of MIDAS (Manchester Information Datasets and Associated Services) has introduced Casweb, a web-based interface to explore the 1991 UK Census area statistics (*URL 3.20*). CDV (Cartographic Data Visualiser) is the software used for the interactive visualisation of the Census data (Dykes, 1998). Regrettably, this possibility for exploration is only available to registered UK academics. However, currently the software Descartes is being tested for the online cartographic exploration of census data. To this end, also non-registered users have access to the software and a Leicestershire Demonstrator Dataset (*URL 3.21*) (also see Andrienko *et al.*, 1999). Further developments may be expected at this side of the map use cube, as the WWW environment is pre-eminently a visual exploration environment.

For all these map use goals, the extremely important question may be posed whether the maps that appear on the display screens during or after a WWW-session really are as efficient and effective as they could be. That is, do the users always get an appropriate answer to the geographical questions they have?

As cartographers always did, web map designers must take into account the purpose of the map and the needs and characteristics of its users (also see Chapter 5). And, in view of the current possibilities for users to produce their own maps as described above, this requirement also holds for the design of the cartographic tools offered to the users, as well as for the design of the user interfaces on the site.

The problem is that we hardly know anything about how people use web maps, or more generally, how people use the WWW to retrieve geographical information. Perhaps we also do not know enough about who is using web maps. The user profile is becoming more and more diversified (see Section 4.1) and we need to know more about the different needs and different characteristics of the different user groups. In any case, the users themselves would certainly be helped if it could be made more clear which websites meet their requirements.

To some extent, the required web map use research is not different from map use research that has already been (and still has to be) executed in other map use environments (van Elzakker & Koussoulakou, 1997). For example, the answers to questions like when, why and how people are using maps in the exploration of geographical data are as much needed in the stand-alone GIS environment as in the WWW environment. Likewise, the results of research into the perception properties of visual variables (including the new "derived" and "dynamic" ones) as applied to cartographic symbols (see Chapters 5 and 7) are relevant in all circumstances in which maps are displayed on monitor screens. And knowing more about the specific backgrounds and characteristics of users which affect their ability to perceive and/or to comprehend the geographical information inherent in the map (e.g. age, previous education, existing knowledge and experience) is not only relevant for the design and development of cartographic tools for the Web.

Some aspects of map use may however be very specific for the WWW-environment and will have to be investigated separately. For example: What are the typical characteristics of the Web search and surf process in which answers are sought (to be given through maps) to various geographical questions? What is the role of the user interfaces in this process? What are the consequences of the volatility of the medium that is used by rather impatient users? And do the web maps generated by the users themselves actually provide the information required, or do they give cause for misinterpretations? Finally, what is the quality and reliability of the geographical information transferred through cartographic displays on the WWW?

In view of the very recent rise of the new medium, it is not surprising that, so far, hardly any web map use research has been executed. As usual, technical developments precede usability questions. However, a start has been made with investigating how maps are being used on the Internet. Examples are the work of Harrower *et al.* (1997) and the extensive customer survey and online user feedback option on the website of the National Atlas of the United States (*URL 3.22*, click the "Atlas Feedback" button) (also see Wright, 1999). Peterson (1997) also mentions the web map use research associated with the development of the Alexandria Digital Library (*URL 3.23*). At this site, map use is being studied by examining the log files of web sessions. These files contain information on the types of maps that are accessed, how long they are viewed, what map is viewed before and after, and where the user clicks on the map. This kind of work should be followed by many more investigations of the use and the users of maps on the Web, so as to be able to develop more effective cartographic tools to better serve the needs of the users. One of the topics to be researched is how people actually find on and retrieve from the WWW the maps and data they need to answer the geographical questions they have.

3.4 FINDING AND RETRIEVING MAPS AND GEODATA ON THE WEB

The WWW is a very rich information source, and the amount of information is growing at an exponential rate. A substantial part of this information is of a geospatial nature. And as maps remain very efficient and effective means for the transfer of this kind of information, they often appear on the monitor screens of users during and after WWW sessions. In these sessions, users may just be browsing the Web for no particular reason (other than mere curiosity) or they may be searching for specific information, e.g. geographical information. Here, we assume that an ever-increasing majority of users starts a web session with an aimed search. As already stated in the introduction to this Chapter, sometimes users did not specifically ask for a map display to appear, but they simply posed a geographical question, "How can I reach Paris from Amsterdam by car?" The answer to such a question may be provided automatically by a more or less customised map (next to a route description in words, generated by one of the routeplanners on the Web (see Chapter 12)). Other users may specifically look on the Web for a particular map of a particular area, because they know or expect that such a map will meet their need for geographic information. In some cases they are even creating the map they want themselves with the help of cartographic functionalities offered at certain websites or with their own cartographic software after retrieving (downloading) geospatial data from the Web.

How do users find the maps or geodata they need on that overwhelming World Wide Web? In fact, the answer to this question is not really known, due to the same lack of web map use research already put forward in Section 3.3 above. The only thing to do here is to present some possible ways of finding and retrieving maps and geodata on the Web and to give utterance to a few hypotheses.

One such hypothesis is, for instance, that users try to find an answer to their geographical questions at websites that were made known to them by other users before and that are revisited time and again. This may lead to an exponential growth of the number of "hits" on such a website, whereas in the meantime perhaps another website exists that may give a more effective answer to the geographical question at hand. It is rather difficult to accept, for instance, that the enormous growth of the number of maps generated on the MapQuest site (see Section 4.2) is brought about solely by the effectiveness of these maps.

However, at the same time it must be realised that it is very hard to find the right map or geodata on the Web. Use can be made of search engines, subject directories, subject guides and specialised databases (Figure 3.3).

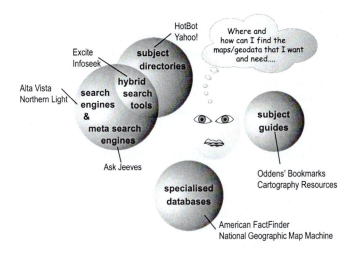

Figure 3.3 Tools for finding maps and geodata on the Web (categorisation derived from Library, University of California, Berkeley, 1999).

Search engines and subject directories are the most common and the most general (trying to cover the Web as a whole). Nowadays, they often are of a hybrid nature (e.g. Excite and Infoseek). Search engines automatically make up indexes of words (and possibly file types) found on web pages with the help of so-called web crawlers (e.g. Alta Vista or Northern Light (*URL 3.24*)). Searching may be done by the user by typing in one or more keywords, like "map" and "Netherlands". One problem with these search engines is a long-standing cartographic issue as well: the spelling of geographical names. Not only is there a danger of not finding the required maps or geospatial data because of typing errors but often there is also disagreement about the actual spelling of geographical names, for example when they are transcribed from another language or in the case of exonyms (Cologne or Köln). Help may be provided by search engines that contain a thesaurus, but this particular problem may also be partly solved with the help of clickable index maps used as search tools (also see Section 3.5). Another way out is to make use of human-indexed subject directories that hierarchically list websites by category: for example, in the HotBot directory (*URL 3.25*) there is a subcategory "Maps" in the main category "Reference". Clicking on "Maps" reveals another list of sub-subcategories with links to websites where maps can be found. On these websites the available maps are often grouped by geographical unit (e.g. countries), listed alphabetically in a pull-down menu.

A big problem with all of these search engines and subject directories is that they cannot cope with the increase of the amount of information on the Web. Although, of course, the exact figures are changing continuously (see *URL 3.26*), recent research (Lawrence & Giles, 1999) led to the conclusion that the best search engine of that moment could only find 16% of the information publicly available at that time. A year before, the best search engine (another one) could still find 35% of the information. A metasearch, combining the results of several search engines together, could at maximum reach a coverage of 42%. At the same time, search engines are more likely to index the more popular sites that have more links to them and in international search engines there still is a bias towards commercial sites based in the United States. Finally, sometimes it may take months before search engines discover new web pages and this works against the potential advantage of actuality as presented in Section 3.2.

Another problem with search engines is that users may have difficulties in making the best possible search. They may find it difficult to choose the right keywords or to formulate their actual search questions. This is sometimes called the "vocabulary differences problem" (Chen *et al.*, 1998). Search engines may help users to formulate complicated search questions (called "power searching"), and sometimes even allow questions put in natural language (e.g. Ask Jeeves (*URL 3.27*)), but still this does not always lead to finding the right map or geographical information (assuming it exists on the Web). Prospective web map use research may perhaps lead to "specialty" metasearch engines that may help users to formulate their specific geographical questions and find the maps and geodata that may help to answer these questions. Perhaps it will also be made possible then to assist users with assessing the quality of the maps and geodata found (by providing metadata), so that they may be able to make a better choice between the various alternatives offered (also see Section 3.5 on geospatial data clearinghouses).

The lack of specific knowledge of what users with geographical questions are looking for is also a big disadvantage of the general subject directories produced by human editors. These subject directories are first of all suited for general browsing and for the most popular search questions (e.g. Yahoo!)

In that respect, the so-called subject guides may be currently of more interest to web users searching for specific information, for instance maps or geodata. Subject guides are hierarchically organised lists of links to relevant websites usually compiled by experts (individuals or institutes or organisations) in a certain subject domain. Well-known examples in the cartography domain are Oddens' Bookmarks (*URL 3.28*) and the Cartography Resources in the WWW Virtual Library (*URL 3.29*). With a particular way of organising the hyperlinks and a user friendly interface, a cartographic subject guide may even become a "virtual atlas" on the Web (Ashdowne *et al.*, 1997). Subject guides may also be limited to rather narrow subdomains: *URL 3.10*, for instance, also contains links to sites with real-time traffic reports of places all over the world.

Finally, whereas the search tools discussed so far (search engines, subject directories and guides) refer to a number (but not all!) websites where the desired information may be found, users looking for specific maps or geodata may also go directly to a specialised database accessible through the Web (if they know where to find them). There are many such databases containing ready-made maps and/or geodata which may be analysed and manipulated. A very nice example is the American FactFinder (*URL 3.30*) which allows the user to search, browse, retrieve, cartographically visualise, print, save and download geodata from many U.S. Census Bureau sources. It all depends on the client-server architecture (see Chapter 6) whether the required information is retrieved from the database as maps or as geodata which are downloaded to the computer of the user (and, perhaps, visualised with the help of cartographic software installed locally). When maps are retrieved as pictures it also depends on the technical set-up whether the maps may be customised (in the GIS at the server side) or whether the user will have to make do with fixed maps that cannot be manipulated at all. Examples include the National Geographic Map Machine (*URL 3.12*) and the PCL Map Collection (*URL 3.13*) respectively.

Examples of websites through which you can download geodata from a remote database (perhaps after browsing and selecting the available data with the help of certain map tools) are the Alexandria Digital Library (*URL 3.23*) and ArcData Online (*URL 3.31*). Particularly with geodata sites it may happen that users do not have free access, either because of reasons of privacy or of copyright (e.g. the Casweb site (*URL 3.20*) already discussed in the previous section). At commercial data sites, users can only download the data if they actually pay for them (see section 4.3). If official geodata are a public good, like in the United States, they may be downloaded free of charge. In the next section, special attention will be paid to recent international developments related to the organisation, availability and accessibility of geospatial data on the Web.

3.5 WEB MAPS IN GEOSPATIAL DATA INFRASTRUCTURES

Because of technological developments in the field of geospatial data acquisition (think, for instance, of satellite remote sensing), everywhere in the world we have witnessed an enormous increase in the amount of available geodata. In recent years the accessibility of up-to-date geodata has improved considerably, due to the increasing maturity of the Internet and the WWW in particular. At the same time, developments in computing technology have made it possible to process unprecedented amounts of geodata very quickly and developments in society have forced organisations and commercial companies that produce maps and geodata to do that in an efficient and cost effective way. All this has led to a growing perceived need to share and integrate geodata resources and geodata production processes, next to the requirement to make it possible for the users to find easily and to retrieve the geodata they need (see Section 3.4).

In recent years, therefore, there has been a growing awareness of the need for geospatial data infrastructures. Usually, these are conceived at a national (sometimes regional) level. But the WWW in particular is destroying boundaries, not only between data producers and data users, but also between the geographical units to which the data pertain. As already reflected by its name, the World Wide Web leads to globalisation, and initiatives are already taken to arrive at a Global Spatial Data Infrastructure (*URL 3.32*). On the other hand, policy and copyright problems seem to stand in the way of the establishment of a European geospatial data infrastructure (*URL 3.33*) and, perhaps, many national spatial data infrastructures (nsdi). As yet, there are only a few countries in the world (among them the United States and Canada) where a national geospatial data infrastructure seems to be more than wishful thinking. See *URL 3.34* for an overview of websites related to local, national and international geospatial data infrastructures.

When talking about a geospatial data infrastructure, reference is not just made to the physical transportation of geospatial data over the Internet. The U.S. Federal Geographic Data Committee (1999) states in a concise way that it:

"... encompasses policies, standards, and procedures for organizations to cooperatively produce and share geographic data."

As such it is a cooperation between governments, the academic community and the private sector. National Atlases (see Chapter 10) and National Mapping Organisations (see Chapter 8) may, for instance, participate in national geospatial data infrastructures, but also commercial companies that maintain, for instance, road databases or offer Internet geotechnology. The geospatial data infrastructure GeoConnections Canada sets a fine example (*URL 3.35*). For the users of geodata, such a cooperation is advantageous, as it leads to geodatasets that may be better integrated and have a known quality. But perhaps the biggest advantage for the users is the improved accessibility of the data in the sense that it will be easier for them to find the geographic information they actually need. In this context, so-called geospatial data clearinghouses will play a prominent role. CEONet, for instance, is the clearinghouse for the Canadian Geospatial Data Infrastructure (*URL 3.36*). The idea is not new (in the 1980s (Bakker *et al.*, 1987) clearinghouses were already being discussed during the development of a new concept for the National Atlas of the Netherlands) but it acquired a new meaning as an important element of a geospatial data infrastructure in a WWW environment. In this new sense, a clearinghouse is a website that may be considered as the "market place" where producers and users of geospatial data meet each other. The clearinghouse website contains field-level descriptions of geospatial data available in the infrastructure (FGDC, 1999). This descriptive information, known as metadata, is collected in a standard format to facilitate querying by the users. As such, the metadata also supply the necessary information on quality (see section 3.4). Clearinghouse sites may provide hypertext linkages to data producers that enable users to directly download digital geodata. Where data are not in a digital form, or where digital data are too voluminous to be made available through the Internet, or the data products are made available for sale, linkage to an order form can be provided in lieu of a data set (FGDC, 1999) (also see Section 4.3).

What role do web maps play in these clearinghouses and geospatial data infrastructures? First of all, as stated before in this chapter, the geodata retrieved may come to the user in map form. But, in addition, web maps may be useful to visitors of clearinghouse sites by fulfilling two other functions:

* web maps as (part of) the searching mechanism;
* web maps as preview of the geospatial data to be downloaded.

When searching for geodata on a clearinghouse site, users will normally be asked to specify things like subject, time period, scale and geographical coverage. This may be done by typing in characters or numbers, or by clicking on the alternatives offered in pull-down menus. However, as far as the geographical area to be covered is concerned, it is often more helpful to define it on a map display of the area the clearinghouse pertains to. The user may do this by clicking on such a map interface, or by drawing a box around the area in which he/she is interested (see e.g. *URL 3.37* and Figure 3.5). Technically, it is even possible to let the user draw a polygon around the geographical area of interest (*URL 3.38*). As a result of such a search action, the clearinghouse (acting like a search engine) may come up with an overview of the geodata available. Users may then be offered to preview these data in map form, before they actually download them. Sometimes, special data viewer software is made available to the user to generate these cartographic previews (e.g. ArcExplorer (*URL 3.31*) or Figure 3.4).

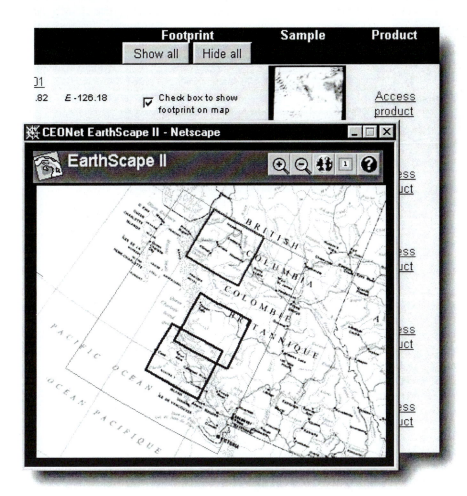

Figure 3.4 Web map preview of geospatial data before downloading (*URL 3.36*).

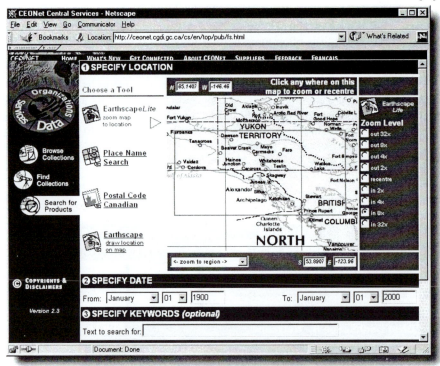

Figure 3.5 Web map search tool in the Canadian CEONet (*URL 3.36*).

URLs

URL 3.1   The website of ITC's Division of Geoinformatics, Cartography and Visualisation <http://www.itc.nl/carto/>

URL 3.2   Non-geospatial category map of entertainment <http://ai2.bpa.arizona.edu/ent/entertain1/>

URL 3.3   Directory of German WWW.ervers <http://entry.de/>

URL 3.4   Dutch Yellow pages <http://www.goudengids.nl/>

URL 3.5   Microsoft's Expedia Maps <http://maps.expedia.com/>

URL 3.6   Historical maps Bremen University <http://gauss.suub.uni-bremen.de/>

URL 3.7   Radar simulation precipitation in the Netherlands <http://weerkamer.nl/radar>

URL 3.8   Latest Atlantic hurricane information <http://hurricanes99.com/>

URL 3.9   Real-time traffic congestion map of Athens <http://www.transport.ntua.gr/map/>

URL 3.10 Predicted traffic flow in Duisburg <http://traffic.comphys.uni-duisburg.de/>

URL 3.11 Real-time positions of sailing-boats <http://www.roundtexel.com/>

URL 3.12 National Geographic's Map Machine <http://plasma.nationalgeographic.com/mapmachine/>

URL 3.13 PCL Map Collection <http://www.lib.utexas.edu/Libs/PCL/Map_collection>

URL 3.14 Deaths from cholera in London, 19th July to 2nd October 1866
    <http://www.geog.qmw.ac.uk/gbhgis/gisruk98/index.html#cholera>

URL 3.15 Make your own map <http://www.aquarius.geomar.de/omc/make_map.html>

URL 3.16 Limited interactivity <http://www.lonelyplanet.com.au/dest/dest.htm>

URL 3.17 Dutch High-Speed Line Project
    <http://www.hslzuid.nl/hsl/uk/intro-uk.html>

URL 3.18 Lycos World Atlas <http://versaware.atlaszone.lycos.com/>

URL 3.19 CIESIN Demographic Data Viewer
    <http://plue.sedac.ciesin.org/plue/ddviewer/>

URL 3.20 Casweb of MIDAS <http://census.ac.uk/casweb/>

URL 3.21 Exploring Leicester Census data with Descartes
    <http://lenny.mcc.ac.uk/kindsdb6/> or
    <http://kandinsky.mcc.ac.uk/VisGat/Descartes/testDescartes.asp>

URL 3.22 National Atlas of the USA <http://www.nationalatlas.gov/>

URL 3.23 Alexandria Digital Library <http://www.alexandria.ucsb.edu/adl.html>

URL 3.24 Northern Light search engine <http://www.northernlight.com/>

URL 3.25 HotBot search engine <http://www.hotbot.com/>

URL 3.26 Search Engine Watch, also with web searching tips
    <http://searchenginewatch.com>

URL 3.27 Ask Jeeves search engine <http://www.askjeeves.com>

URL 3.28 Oddens' Bookmarks <http://oddens.geog.uu.nl/index.html>

URL 3.29 Cartography Resources on the Web
    <http://geog.gmu.edu/projects/maps/cartogrefs.html>

URL 3.30 The American Factfinder website of the U.S. Census Bureau
    <http://factfinder.census.gov/java_prod/dads.ui.homePage.HomePage>

URL 3.31 ArcData Online geodata source
    <http://www.esri.com/data/online/index.html>

URL 3.32 Global Spatial Data Infrastructure homepage <http://www.gsdi.org/>

URL 3.33 MEGRIN: a first step towards a European Geospatial Data
    Infrastructure? <http://www.megrin.org/index.html>

URL 3.34 Links to sites related to geospatial data infrastructures
    <http://www.gsdi.org/sdi.html>

URL 3.35 Canadian Geographical Data Infrastructure GeoConnections
    <http://cgdi.gc.ca/>

URL 3.36 CEONet clearinghouse (Canada) <http://ceonet.cgdi.gc.ca/>

URL 3.37 U.S. National Spatial Data Clearinghouse custom search form with map
    as search tool <http://fgdclearhs.er.usgs.gov/customsearch.html>

URL 3.38 Looking for a house in Germany? Draw a polygon around the area of
    interest ("Erweiterte Suche") <http://www2.rdm.de/cgi-bin/rdmscript.cgi>

REFERENCES

Andrienko, G., Andrienko, N. and Carter, J., 1999, Thematic mapping in the
    Internet: exploring Census data with Descartes. In *Proceedings TeleGeo'99,
    Lyon, France*, edited by Laurini, R., (Lyon: Claude Bernard University of Lyon),
    pp. 138-145.

Ashdowne, S., Cartwright, W. and Nevile, L., 1997, A virtual atlas on the World
    Wide Web: concept, development and implementation. In *Proceedings, Volume
    2, of the 18th International Cartographic Conference ICC97, Stockholm*, edited
    by Ottoson, L., (Gävle: Swedish Cartographic Society), pp. 663-672.
Bakker, N.J., Elzakker, C.P.J.M. van and Ormeling, F.J., 1987, National atlases
    and development. *ITC Journal*, (1), pp. 83-92.
Baumann, J., 1999, Geotechnology spices up France's premier yacht race.
    Geofeature: GIS / Web-based Mapping. *GEOEurope*, **8** (4), pp. 38-39.
Chen, H., Houston, A.L., Sewell, R.R. and Schatz, B.R., 1998, Internet browsing
    and searching: user evaluations of category map and concept space techniques.
    Journal of the American Society for Information Science, **49** (7), pp. 582-603.
Dykes, J.A., 1998, Cartographic visualization: exploratory spatial data analysis
    with local indicators of spatial association using Tcl/Tk and CDV. *The
    Statistician*, **47** (3), pp. 485-97.
Elzakker, C.P.J.M. van and Koussoulakou, A., 1997, Maps and their use on the
    Internet. In *Proceedings, Volume 2, of the 18th International Cartographic
    Conference ICC97, Stockholm*, edited by Ottoson, L., (Gävle: Swedish
    Cartographic Society), pp. 620-627.
FGDC (1999), Home page Federal Geographic Data Committee.
    <http://fgdc.er.usgs.gov/index.html> (accessed 11.11.1999).
Harrower, M., Keller, C.P. and Hocking, D., 1997, Cartography on the Internet:
    thoughts and a preliminary user survey. *Cartographic Perspectives*, (26), pp.
    27-37.
Krygier, J., 1999, World Wide Web mapping and GIS: an application for public
    participation. *Cartographic Perspectives*, (33), pp. 66-67.
Lawrence, S. and Giles, C.L., 1999, Accessibility and distribution of information
    on the Web. *Nature*, (400), pp. 107-109.
Library, University of California, Berkeley, 1999, Important things to know before
    you begin searching the Web. *Teaching Library Internet Workshop*,
    <http://www.lib.berkeley.edu/TeachingLib/Guides/Internet/ThingsToKnow.html>
    (accessed 7.11.1999).
MacEachren, A.M., 1994, Visualization in modern cartography: setting the agenda.
    In *Visualization in modern cartography. Modern Cartography, Volume Two*,
    edited by MacEachren, A.M. and Taylor, D.R.F., (Oxford: Elsevier Science Ltd.
    / Pergamon), Chapter 1, pp. 1-12.
MacEachren, A.M. and Kraak, M.J., 1997, Exploratory cartographic visualization:
    advancing the agenda. *Computers & Geosciences*, **23** (4), pp. 335-344.
Peterson, M.P., 1997, Cartography and the Internet: introduction and research
    agenda. *Cartographic Perspectives*, (26), pp. 3-12.
Wright, B., 1999, The National Atlas of the United States of America. In
    *Proceedings of the Seminar on Electronic Atlases and National Atlas
    Information Systems in the Information Age, held at the University of Iceland
    (Reykjavik), 1998*, edited by Gylfason, A., Köbben, B., Ormeling, F.J. and
    Trainor, T., (ICA Commission on National and Regional Atlases), pp. 35-40.

CHAPTER FOUR

Users of maps on the Web

Corné P.J.M. van Elzakker

4.1 USER PROFILES

As demonstrated in Section 3.3, there is a need to know more about who is using which web maps for what purpose. This need is becoming more and more pressing as the population of users as well as the map use goals are diverging rapidly nowadays. Three years ago we also did not know much about the use and users of web maps. However, we did know that the group of people who actually made use of the Internet was not very diverse at that time. Therefore, three years ago it was possible to state (van Elzakker & Koussoulakou, 1997) that the group of users of maps on the WWW could be defined pretty well: relatively young (15 to 40 years of age) males in Western countries with a high level of education, with an interest in science, technology and/or computers and in the possession of a PC. Also in view of the specific characteristics of the WWW medium, they were sometimes considered as a completely new generation of map users who were and are interacting with map displays in entirely different ways than "traditional" map users. But still, because of the rather limited group of people actually connected to the Internet, it was not so difficult in theory to identify web map purposes and to adjust the cartographic web tools to the needs and characteristics of its potential users.

However, in the meantime things have changed, and still are changing, rapidly. As with all developments related to the WWW on the Internet, North America has taken the lead and demonstrates a significant change of the web user profile. More and more information is being collected on this profile because of growing commercial interests in it (see Section 4.3). User data are made available through several websites (e.g. *URLs 4.1, 4.2* and *4.3*). The Internet now plays a role at all levels of education and is becoming more and more common in every home and business. In the United States, most users now access the Web primarily from home, whereas they primarily did it from work in the early days (Kehoe *et al.*, 1999). Peterson (1999) reports on an investigation among people planning to get Internet access: almost half of them only have a high school education or less; 58% of them make less than $50,000 a year. The use of the Internet is democratising, although a deep gap may come into existence with the lower social classes that do not move on to the information highway. International Data Corp. (IDC) expects that 62% of all adults in the United States will have Internet access by 2003 (CyberAtlas, 1999a). In mid-1999 in the US there were clear signs of an overtaking manoeuvre by women: already 47% of the users were female. At the same time, 22% of all online adults were 50 years or older (Cyber Dialogue, 1999). For older people the advantage of accessibility (as discussed in Section 3.2) is perhaps relatively even more important than for younger people.

At least part of the world will undoubtedly follow the American example and demonstrate similar changes in user profile in the years to come. For instance, many European countries are catching up rapidly. As a consequence, there will be more and more different web map users with different needs and requirements. Some of these potential web map users may be regarded as "new" users, in the sense that maps now are much more accessible to them, and before they would

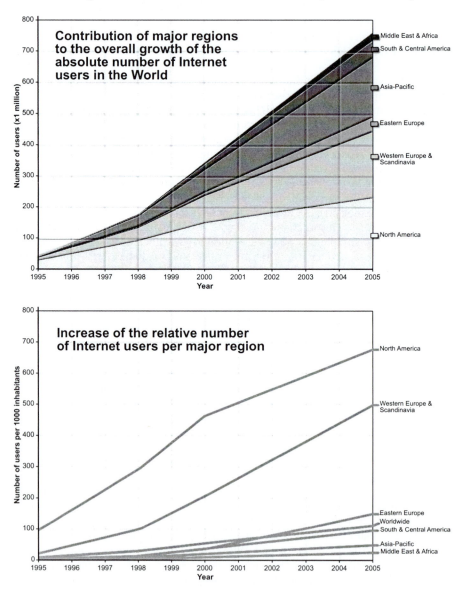

Figure 4.1 Worldwide Internet users 1995-2005 (source: *URL 4.4*, December 1999).

normally not consider buying such a thing as GIS software. The Internet will make it possible for them to really interact with maps for the first time, so that all kinds of individual geographical problems may be solved much more efficiently and effectively than ever before. All this means that more and more attention should now be paid to adjusting the cartographic websites to specific user groups. For instance, the nature of the user interface and the possibilities for interaction cannot be the same for primary school children and for geoscientists exploring a geospatial dataset.

It is not only a matter of a user profile that is becoming more and more diversified. At the same time, the number of users of the Internet and people with access to the WWW is still growing exponentially. Estimates have been made by several organisations and have to be readjusted all the time. A November 1999 estimate by Computer Industry Almanac, Inc. arrived at 259 million Internet users for year-end 1999 (CyberAtlas, 1999b). Internet users are defined here as adults over 16 years old with weekly usage in business and homes. The numbers are said to be 15 to 30% higher when occasional Internet users are included. Supposedly, the numbers would also increase if the (rapidly growing) use at schools and, for instance, in public libraries and cybercafés would be included. In any case, the number of users will grow rapidly in the years to come (see Figure 4.1). At the same time, however, it should be realised that the number of Internet users expressed as a percentage of the total population of the world (see Figure 4.2) will still be rather small, even five years hence. This is mainly a matter of the global diffusion of the Internet. By year-end 1999, for instance, 43% of the world total of Internet users lived in the United States. This figure will decline to 27% by the end of 2005 (CyberAtlas, 1999b).

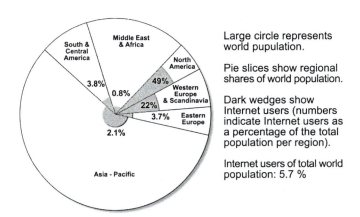

Figure 4.2 The portion of the world's population that uses the Internet, forecast year-end 2000 (source: *URL 4.4*, December 1999) (diagram conceptualised by UNDP, 1999).

Indeed, certainly in a book dealing with web maps and the dissemination of geospatial data, we do not only want to know who the users are and how many there are. Of course, we also want to know where they are. Figures 4.1 and 4.2 already give an indication of the (development of the) number of users in the major regions in the world. In absolute terms, North America will remain the leading region for Internet users in the years to come, but the other regions are growing faster. In some of these other regions very interesting developments are taking place, like the Cyberjaya project in Malaysia, stimulated by Prime Minister Mahathir. Cyberjaya is a digital city with a so-called e-government that uses no paper and exchanges all information through the Internet (*URL 4.5*).

Figure 1.1 already showed the percentage of population with access to the Internet by country. In this map the different surface areas of the territorial units (in this case countries) have an unwanted effect on the perception of the global diffusion of the use of the Internet (larger countries tend to dominate, although their surface areas are not related to population numbers). Therefore, Figure 4.3 is included here to show the absolute numbers of Internet users by country (using the same data sources as for Figure 1.1). The top 15 nations with the most Internet users at the end of 1999 are represented by means of separate proportional circles. All other countries (with less than 2.5 million Internet users) are put into classes. Together, the Top 15 nations account for nearly 82% of the worldwide Internet users (CyberAtlas, 1999b). The website accompanying this book will be kept up to date on the rapidly changing numbers.

Over 200 nations are connected now, but when considering Figure 4.3, the uneven distribution of Internet (and, consequently, web map) users is striking. Factors that are mentioned (e.g. by Hargittai, 1999) to explain this uneven distribution are: economic wealth, level of education, (English) language proficiency, government policies (e.g. political or religious freedom, freedom of competition leading to differences in Internet access pricing) and existing telecommunication, computing and power facilities. Looking at Africa as a whole, for instance, the low number of Internet users is not very surprising, if only because of the low literacy rates and the low number of fixed telecommunication (telephone) connections. And it should also be realised that within countries the Internet connectivity is often limited to one or two large cities (Press *et al.*, 1999). It is sometimes argued, therefore, that the globalisation that is partly brought about by the Internet is élite-based and at the same time leads to increasing global (and social) inequality (UNDP, 1999). On the other hand, every country in Africa apart from Eritrea was connected to the Internet in November 1999 (*URL 4.6*). Technical progress, in particular a rapid introduction of less vulnerable wireless means of telecommunication, may mean that the dissemination of maps and geospatial data through the WWW could contribute to the further development of this part of the world as well. In some African countries there are already more mobile telephones than fixed telephone connections and soon it will be possible to have access to the Web through a mobile Internet (see Section 4.4)(Stähler, 1999).

In Europe, as in Africa, the penetration of the Internet is very unequal at the moment. The contrast between Western and Eastern Europe (see Figure 4.1) may not be surprising. There is also, however, a contrast between Northern Europe, where the Scandinavian countries Sweden, Norway, Finland and Iceland all have more than 30% of the population with access to the Internet, and Southern Europe, where the Mediterranean countries (e.g. Greece, Italy, Spain and Portugal) have less than 10% of the population using the Internet (see Figure 1.1). The relatively small numbers of Internet users in France and Germany are also striking. It is one of the main reasons, for instance, why the new German national atlas will only appear in print and in an off-line electronic edition and not on the WWW

(Lambrecht & Tzschaschel, 1999). France may be suffering now from its early advance in relation to the introduction of the information system Minitel, originally based on the television network (videotex) and set up before the Internet and the WWW came into being. The fact that English is the dominant language on the Web may play a role here as well.

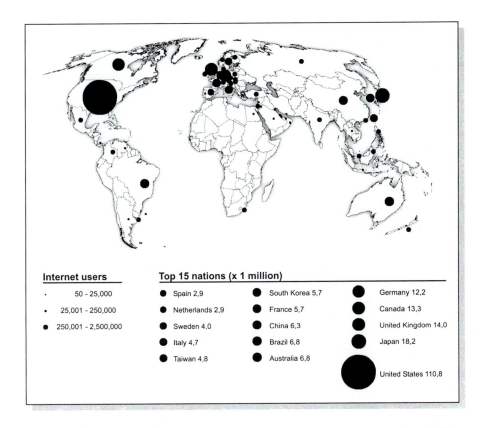

Figure 4.3 Number of Internet users by country, 1999 (sources: *URL 1.8* and CyberAtlas, 1999b).

In the Asia-Pacific region, Japan, Australia, China, South Korea and Taiwan already belong to the Top 15 nations in Internet use (Figure 4.3). The growth of the number of people with Internet access in a country like China is exponential. Depending on the Internet policy of the government and in view of the enormous amount of people living in this country (1.2 thousand million), China may go up in the Top 15 rapidly. And this will also have consequences for the number of maps generated through the WWW.

4.2 HOW MANY WEB MAPS AND WEB MAP USERS?

In the past decades the need for maps has increased enormously in all parts of the world, as a consequence of, among other things, the growing number of geospatial relationships, greater human mobility and more physical planning problems brought about by a more intensive use of land and water. It is a question what influence the exponential growth of the Internet, the corresponding increase of the time people will and may spend at their network stations and the resulting globalisation will have on the intensity of the geospatial relationships and the related need for maps and geodata. In any case there will be a need for maps of cyberspace (see Section 2.3). And because of a better accessibility and actuality of web maps the medium may also generate a greater interest in maps of all kinds (even paper maps!). It should be noted that with the introduction of the Internet and the WWW we have witnessed an enormous increase of the number of maps that are actually produced and used. In quantitative terms, the Web now has become the major medium for the dissemination of maps to their users.

In the previous section estimates were provided on total numbers of Internet users. Many data are also collected on the numbers of "hits" on websites, for these data are used to attract advertisers or to assess how much a website can charge for advertising banners. The only problem is that these data are not made available easily because of the competition between commercial websites and because the data have now become a property, handled by separate and independent companies (Peterson, 1999). At the same time, there are many web map sites that do not advertise and do not keep a record of the number of people that access their site, or use their maps (Peterson, 1997).

Media Metrix keeps up some rankings of websites that are hit by most users (*URL 4.7*). The rankings show the actual number of total users who visited the website once in a given month, whereby all unique visitors are unduplicated (only counted once). Not surprisingly, search engines, the sites of Microsoft and sites like Amazon.com figure high in the rankings. The highest specific web map site, MapQuest (*URL 4.8*), was listed as number 38 on the March 2000 ranking with 5,572,000 different users (compared to rank 49 in November 1999 with 3,754,000 users). However, rankings like these do not give an indication of the total number of web map users nor of the total number of web maps actually retrieved, generated or downloaded. For instance, maps are also an important means of information dissemination on The Weather Channel (*URL 4.9*), figuring as number 27 on the Media Metrix ranking with 7,598,000 different users in March 2000 (November 1999: rank 39 with 4,677,000 users). But we do not know how many of these users actually used maps to get information on the weather, how often they come back to the site, how many maps they used each time and definitely also not how effective the maps were in providing the wanted information.

In October 1998, only 10.4% of the 3291 respondents of a WWW user survey (Kehoe *et al.*, 1999) said that they never looked for a web map; 41.4% accessed maps less than once a month; 32.2% monthly; 14.7% weekly and only 1.3% daily. Assessing the total absolute number of web maps that is produced and used is a very difficult task for reasons mentioned above. However, useful data are available for some specific websites. And these data are sometimes very impressive. For example, as already mentioned in Chapter 3, in 1998 over 5 million web maps were interactively and dynamically created during 10 days of the Route du Rhum sailing race (Baumann, 1999). On average, there were 200,000 hits per day on the race's website. So, on average 2.5 maps were generated during each WWW session. MapQuest (*URL 4.8*) is consistently mentioned as the number one web map site in the world, or, as Crampton (1998) states, the biggest map maker in history. According to a MapQuest employee there were 75.4 million maps drawn on the MapQuest site in November 1999 (Gebb, 1999). These are 2.5 million maps a day or 1,750 maps a minute on average (and it will be much more during peak hours). In November 1999 the MapQuest site had 16.6 million user sessions (cf. the figure of 3.7 million *different* users counted by Media Metrix, as mentioned above). And this means that, on average, some 4.5 maps were generated during a user session. MapQuest is a very popular site indeed, offering various functionalities and a lot of useful geographic information (see Chapter 9 and Figure 9.2). However, perhaps the speed and ease of information retrieval are at the expense of the quality of the cartographic design of MapQuest maps. As a consequence, there may even be doubts about their effectiveness. Following some of the guidelines presented in Part III of this book may lead to web map designs that are better adjusted to the needs and characteristics of their users. And, in turn, this may lead to even higher hit rates and another stimulus for a growing overall popularity of maps as carriers of geographic information over the Web.

On the basis of information derived from selected sample sites like these, Peterson (1999) estimates that approximately 40 million web maps in total are used per day world-wide. This is a four-fold increase of the estimate he made in 1997. An even more dramatic growth of web map usage may be expected as a consequence of the predicted exponential growth of the overall number of Internet users (see Section 4.1). In view of the enormous amount of web maps produced and the expected growth thereof, web cartography cannot but have important economic implications as well.

4.3 ECONOMIC ASPECTS

The Internet as a whole is becoming an important player in the world economy. According to the University of Texas' Center for research in Electronic Commerce the global Internet economy accounts for 2.3 million jobs and a turnover of $507 thousand million in 1999 (*URL 4.10*). Most likely, the Internet is also responsible for more than half of the growth of the United States' Gross Domestic Product in 1999 (the growth of the U.S. GDP is projected to be $340 thousand million). And it is predicted (CyberAtlas, 1999a) that in the U.S. the Internet will account for 7% of the GDP by 2003. We do not yet have economic data on web cartography, but this part of the Internet economy will also take its share of the growth, both in terms of employment and revenues. The Web is not just a means to advertise traditional cartographic products like paper maps or CD-ROMs or map making software (see e.g. *URL 4.11*). Creating (possibilities for) maps on websites (e.g. the Dutch Yellow Pages, *URL 3.4*) is becoming a new specialisation of cartographers and cartographic companies now have to adapt to making available their cartographic products and services through the WWW.

In doing so, these companies are confronted with another kind of economy, that is often referred to as the "Internet Economy" or the "New Economy" (Thoen, 1999). It is an economy that is based on immaterial production and consumption and to quite a large extent currently also on future expectations ("the more web surfers, the better") and on the provision of information goods (including maps and geodata) free of charge. In order to be able to survive, cartographic companies and organisations will have to adjust to the rules of the Internet economy and that will not always be that easy.

However, this chapter is written from the perspective of the web map user and not the producer. For the user, the economic aspects of web cartography may be categorised under the following headings:

- For free or for fee?
- Privacy and security
- Copyrights

As stated before, currently many web maps and much geodata are available to users free of charge and many users consider that to be normal. For instance, next to the CEONet clearinghouse of GeoConnections Canada (see Section 3.4) there is another mechanism, called GeoGratis (*URL 4.12*), that delivers geospatial data free of charge, using ftp (file transfer protocol). The fact that users may expect web maps and geodata to be free of charge has many consequences for the suppliers. Companies that have invested a lot in collecting and processing geodata may have to earn that money back, at least. It should be noted that in a situation like this it will also be more and more difficult, for instance, still to make money from route planners on CD-ROM if there are many free route planners available on the Web (see Section 12).

Several reasons may be brought forward to explain the practice of offering information through the WWW free of charge. The first reason is that the Internet and the WWW did not start as commercial undertakings (Peterson, 1999). In the early days, facilitating the exchange and sharing of information between researchers was the main objective. And when the WWW became at first dominated by the United States it was also used to disseminate government held information, including geodata and web maps, that is considered to be a public good that should be made available to the people free of charge. This U.S. Federal open access policy is different from the cost-recovery policies of many governments in, for instance, Europe and forms part of the explanation why geospatial data infrastructures are more of a reality in the U.S. than in Europe (see Chapter 3). It should also be realised that many European governments are forced to cut costs, improve efficiency and generate more revenues. Another reason for making it possible to disseminate web maps and geodata free of charge is that many websites contain banners with commercial advertisements. Websites with maps and geodata may have very high daily hit rates (e.g. the sites of MapQuest and The Weather Channel discussed in Section 4.2) and, therefore, they are of great interest to advertisers. One way of advertising on the MapQuest site (*URL 4.8*) is by means of showing the locations of, for instance, the hotels of a particular chain along the routes planned. On the other hand, online advertising is generally unpopular with users (see e.g. *URL 4.3*). Finally, companies that are selling GIS or cartographic software may advertise their products by making available free of charge geodata that work with this software (e.g. *URL 4.13*). And companies that sell cartographic products on paper (e.g. atlases) or on CD-ROM (e.g. route planners) may give stripped and free demonstrations of these products through websites by way of advertisement.

Of course, users benefit by having access to free web maps and geodata. However, caution should be exercised. Users may become less willing to pay for web maps and geodata and this may lead to a loss of quality, i.e. a loss of effectiveness, or even a loss of actuality, one of the very advantages of the WWW medium (see Section 3.2). This is because the producers cannot or do not want to make such high initial investments anymore. In traditional cartography, this has occurred, for instance, with paper road maps that are offered to petrol station customers free of charge or very cheaply. One aspect of this problem is that users are not always fully aware of the quality of maps; as long as the map looks nice and detailed, they often think it is their fault if they cannot find their way with it easily. Another reason for users to be cautious of free web maps and geodata is that it also happens that owners of websites make money by selling the personal data they collect from WWW users, with all the related problems of privacy and security.

At the moment, the Web is going through a process of commercialisation. And although many web maps and geodata may still be retrieved free of charge, there are more and more so-called e-commerce sites where users have to pay for the information to be obtained. After all, maps and geodata are commodities that have a value (that can be added in various processing steps), if only because of the costs of their acquisition and production. In its simplest form, websites may be used as an easily accessible shop window for cartographic products and geodata that will not be disseminated to users through the WWW. However, sometimes these cartographic products and geodata may be ordered online. For instance, the Travel Book Shop (*URL 4.14*) is a website for ordering paper maps of countries and regions all over the world. Clickable index maps are used on this site to find the maps in which users may be interested. The paper maps ordered are then sent to the users by ordinary mail services. However, it is also possible that web maps and geodata for which users have to pay are downloaded from the website as digital files. For instance, users have to pay for some of the data that may be downloaded from the ArcData Online geodata source referred to above (*URL 4.13*).

When maps and geodata are ordered or downloaded online, there are several possibilities for the user to pay for these (see Figure 4.4). Next to online payment in advance, payments by means of traditional banking methods or cash on delivery are obvious solutions for maps or geodata that are not disseminated through the Web. For online information retrieval, users may subscribe to certain information services through the Web and access these websites through passwords (e.g. *URL 4.15*), or they may have to pay for individual information products directly and online ("pay-per-use"). This may be done with the help of an identification system (e.g. the new Pentium III processors each have a unique number) and by inserting smart cards in a special smart card reader that is connected to the computer. Information suppliers may also make money out of the use of websites by applying the so-called "call switch" principle: users who want to retrieve information are automatically connected through to another telephone line for which they have to pay more. For the users, it is important, of course, to know how much more.

Online payment may also be executed by providing credit card details to the supplier, as is common practice in Northern America. However, many web users in Europe still have their hesitations for reasons of privacy and security. And indeed it is possible for unauthorised persons to intercept and abuse credit card or other personal details sent over the Internet. Besides, not all suppliers appear to be bona fide. Therefore, all over the world banks (afraid of losing part of their traditional involvement in the traffic of payments), credit card companies and others are working on methods for secure payment. There are ways of encrypting credit card details and sending them through a secure server and a system like SSL (Secure Sockets Layer), that is available with the Explorer and Netscape browsers. This ensures a safe Internet connection between user and supplier (indicated by a closed lock at the bottom of the web page). An even better protection for the user is provided by the international safety protocol SET (Secure Electronic Transaction, *URL 4.16*) that will be used, for instance, in the Dutch online payment system I-pay (*URL 4.17*). SET software also safeguards the user against untrustworthy owners of websites. The main problem for information vendors now seems to be to convince users that online payment is as safe as other modern financial transaction methods.

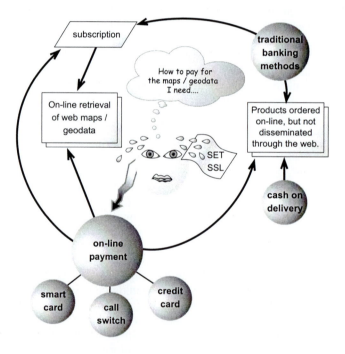

Figure 4.4 Paying for web maps and geodata.

For some users of web maps and geodata, for instance those who want to add value to them, it is also important to know whether the maps and geodata are protected by copyright. Companies (or government organisations in Europe, for instance) that disseminate maps and geodata through the Web may have to do this in order to be able to get something back from high investments made in geospatial data collection and further processing. Legal problems concerning copyright on the WWW are not really solved up to now. The copyright of web maps may however be reinforced by special techniques like watermarking (e.g. *URL 4.18*), by displaying the map extracts in special shapes and sizes, or even by restricting the access to the web maps or geodata to those with a special entitlement or licence. A website exists (*URL 4.19*) that examines several general issues of copyright that also apply to web maps (Crampton, 1995). It should be noted, however, that this website is mainly based on U.S. copyright laws and these may be different from the laws in other countries.

The general conclusion from the above discussion is that at least for some users there are some economic problems related to the retrieval of maps and geodata from the WWW. There are some other problems and limitations as well, as will be discussed in the last section of this chapter.

4.4 PROBLEMS AND LIMITATIONS

For users, some limitations of web maps are not directly related to the WWW, but are a consequence of the computer nature of the medium (as opposed to traditional paper maps), like limited portability and lack of ease in manipulating the map (folding, turning, drawing or measuring on it) and limited display size. The latter may be overcome by options such as panning or zooming, but the limited possibilities for overview are still a disadvantage. In addition, screen and colour resolution usually limit the amount of detail present on a monitor screen map. A paper map will have a higher resolution and probably more detail and a higher information density at the same scale (van Elzakker & Koussoulakou, 1997). In these respects, paper media such as atlases still have some inherent advantages (see Chapter 10).

Throughout this chapter and Chapter 3 mention has been made of some other user problems, which may, to some extent, be of a general cartographic nature, but which are mentioned here in direct relation to the dissemination of maps and geodata through the WWW. For example, in Section 3.3 reference was made to the problem that not enough is known yet about the effectiveness and efficiency of web maps and the sites in which they are embedded, nor about the needs and characteristics of the web map users (also see Section 4.1). A related problem is the quality of the design of web maps. In Section 4.2, for instance, the quality of the MapQuest maps was called into question. One aspect of the problem is that everyone may now design and construct maps and disseminate them through the WWW, or do that on the basis of geodata obtained through the Web, even without having the necessary cartographic knowledge or background. This implies the risk that cartographic design rules are violated, or not applied to their full potential, with a resulting loss of the effectiveness of web maps. Another aspect of the problem is that the creators of web maps do not have full control over their final appearance. Although web maps are stored in platform-independent formats (e.g. GIF, JPEG or PDF), they do not appear exactly the same for every user. The effects of the cartographic designs may differ greatly depending on the various output configurations used. Even considering PCs only (so not new Internet appliances like set-top boxes for online digital TV), there will be differences in the users' browsers and operating systems (which handle colours in different ways, for instance) and in the quality (e.g. resolution) of their graphic cards and display screens (e.g. LCD or CRT in different sizes). Besides, users are able to personally adjust their displays for things like resolution, contrast, brightness and colour balance. The next part of this book will provide some hints on how to deal with these different output conditions as well as some basic cartographic design rules.

This part of the book started in Chapter 3 by presenting the two main advantages of the WWW medium from the perspective of the web map user: accessibility and actuality (see Section 3.2). It will be clear by now that these advantages are not (always) fully achieved. The problem of some websites is that they are not kept up-to-date regularly. As a consequence, users will lose their confidence in these sites. What is more important is that, in practice, there may be quite some limitations to the accessibility. These limitations may be listed under the following headings:

- Finding web maps and geodata
- Language
- Accessibility for everyone?
- Web maps and geodata for fee
- Internet access
- Speed of data transfer

Section 3.4 dealt with the problems users have when "drinking from the fire hose" (van Elzakker & Koussoulakou, 1997), i.e. in finding the maps or geodata they need on the information-rich WWW. A related problem is the volatility or continuity of the information: what appears in a site today might be gone tomorrow.

Language also plays an important role in the accessibility. Misspelling (e.g. of geographical names) may lead to not finding the required web maps or geodata. And although the Web is not limited by political boundaries, the worldwide dissemination of maps and geodata may be hindered by language problems. English is the dominant language on the WWW (*URL 4.20*: 86.55% of the web pages are in English), but not everyone understands this language (only some 10% of the world's population do so). Besides, it should be realised that some 22% of the world's adult population still is illiterate (UNDP, 1999).

Indeed, in Section 4.1 it is demonstrated that the WWW is not yet accessible to everyone. Even in societies with a literacy rate of (almost) 100% certain social classes do not have access to the information highway and Figures 1.1, 4.1, 4.2 and 4.3 show that there are substantial geographical anomalies too. It seems that currently access is limited to people or areas with a certain economic wealth, a certain level of education and computer skills, (English) language proficiency, a favourable government policy and the necessary equipment. Economic factors alone are perhaps the most important explanation for limitations in web access.

Accessibility is fostered by the availability of many web maps and geodata free of charge. This may lead to problems of quality, as stated in Section 4.3. These problems may perhaps be alleviated by the establishment of geospatial data infrastructures (McGranaghan, 1999). However, in Section 3.5 mention was made of policy and copyright problems that stand in the way of GSDI's and new limitations are coming up now that users have to pay more often for what they retrieve. In this respect, individual users also experience problems with privacy, security and online payments (see Section 4.3).

But getting web maps and geodata free of charge is an illusion anyhow. The first requirement is Internet access, and this now means (if you do not want to rely on, for instance, a public library or a cybercafé where you will also have to pay something for getting access) having a powered computer with a modem, connected to a telecommunications network. Next to this hardware (and some software) the user, or his or her organisation, has to pay for the telephone costs and/or an Internet provider. In some places, e.g. in developing countries, these costs are relatively very high, but in other countries these costs are reduced in order to attract as many new Internet users as possible. It also means that Internet access now still is limited to places with a connection to a (properly functioning) fixed telecommunications network, i.e. at home or at work. Therefore, obtaining maps and geodata through the WWW while one is away from one's base is currently not commonplace.

However, technological developments are very fast and it may be expected that the mobile (wireless) Internet will come to stay within a few years. Several solutions are currently under development (e.g. the Internet through satellites) or already put on the market. An example of the latter is the Internet through mobile telephones (GSM), perhaps in combination with notebook computers, based on WAP (Wireless Application Protocol) (*URL 4.21*, also see Appendix A) and future variants thereof (GPRS and UMTS). The traditional Swiss map making company Kümmerly+Frey AG has set up an interesting project called MOGID (Mobile Geo Information on Demand) that aims at bringing to users maps and geodata where they need them, with the help of palm-sized PTAs (Personal Travel Assistants) or PICs (Personal Intelligent Communicators) with a built-in GPS (Global Positioning System) (Sollberger, 1999). A consequence of technological developments like these is that the size of the graphic display screens will be small (compared to the monitors of PCs at home and at work), and they may be monochrome only. And this forms another challenge for the designers of web maps! Perhaps there will soon be a need for a book on WAP cartography, for instance.

A final current limitation to the accessibility is the speed (and reliability) of data transfer through the Internet. For users, speed is one of the biggest problems in using the Web (Kehoe *et al.*, 1999) and it often still is the very advantage of a medium like CD-ROM for the dissemination of atlases, route planners, maps and geodata. A website that shows the condition and performance of the Internet by means of animated maps of various parts of the world is The Internet Weather Report™ (*URL 4.22*). The animations are based on time sequences, as the time of the day really matters for the speeds of data transfer (are the North Americans sleeping or not?) But, of course, the speed strongly depends as well on the technology available to each user, not just his or her own PC and the speed of the modem, but, for instance, also the capacity of the local telephone, ISDN or cable networks. Web maps and geodata usually come in large files and it may take a long time to retrieve or download them from the Web. Therefore, they are prone to the World Wide *Wait* syndrome of the many users who are rather impatient and unwilling to wait for maps to download. This is the reason for a commercial company like MapQuest (see Section 4.2) to continuously increase its website capacity to be able to deal with the peak hours. For non-commercial owners of websites this will not always be possible (Peterson, 1999). If technology would not change, the problem would become bigger and bigger, because of the exponentially increasing use of the WWW. The reliability and speed of the Internet are however constantly improving and many new technological developments may be expected that will further increase the bandwidths and the speed of data transfer. For instance, broad band xDSL techniques (Digital Subscriber Line, widely applied already in the U.S.) will substantially increase the speed of data transfer through ordinary telephone lines. At the same time, a new parallel Internet 2, based on a fibre-glass network, is currently under development and should provide very fast connections between scientific institutes (*URL 4.23*). Indeed, some people argue that speed is not a technical but in fact an economic problem: the solutions are there, as long as the user wants to pay for them. However, web map designers may provide their contribution by constructing web maps that have a storage size (and therefore a download time) that is as small as possible. This may mean simpler and smaller maps with a smaller graphic and information density than their counterparts disseminated through other media.

It was stated before (see Chapter 1) that there are many static view only maps on the Web that are just scans of original paper maps. But it should now be clear that some of the limitations and user needs mentioned in this part of the book dictate that web maps have special design requirements. The next three chapters of this book deal with the principles and practice of effective, efficient web map design in order to try to overcome or minimise some of the problems and limitations listed here.

URLs

URL 4.1 Matrix Information Directory Services <http://www.mids.org/>

URL 4.2 IDC Project Atlas <http://www.idc.com/>

URL 4.3 UK Internet User Monitor
                 <http://www.fletch.co.uk/content/monitor/method.html>

URL 4.4 Computer Industry Almanac, Inc. Internet users by region
                 <http://www.c-i-a.com/199908iu.htm>

URL 4.5 Cyberjaya project in Malaysia <http://www.cyberjaya-msc.com/>

URL 4.6 African Internet connectivity <http://www3.sn.apc.org/africa/>

URL 4.7 Media Metrix Top 50 <http://www.mediametrix.com/usa/data/thetop.jsp>

URL 4.8 MapQuest <http://www.mapquest.com/>

URL 4.9 The Weather Channel <http://www.weather.com/>

URL 4.10 The Internet Economy Indicators <http://www.InternetIndicators.com/>

URL 4.11 Advertising on the Web <http://www.delorme.com/>

URL 4.12 GeoGratis, another element of Canada's GeoConnections
                 <http://geogratis.cgdi.gc.ca/>

URL 4.13 ArcData Online geodata source
                 <http://www.esri.com/data/online/index.html>

URL 4.14 Travel Book Shop <http://www.travelbookshop.ch/>

URL 4.15 Subscribing to Michelin's travel information services
                 <http://www.michelin-travel.com/home.cgi>

URL 4.16 Secure Electronic Transaction <http://www.setco.org/>

URL 4.17 I-pay <http://www.i-pay.com/uk/merchant/>

URL 4.18 Example of watermarking
        <http://media.maps.com/magellan/Images/NETHER-W1.gif>
URL 4.19 The (U.S.) Copyright Website <http://www.benedict.com/>
URL 4.20 Inktomi WebMap <http://www.inktomi.com/webmap/>
URL 4.21 WAP home page <http://www.wapforum.org/>
URL 4.22 The Internet Weather Report™ <http://www.mids.org/weather/>
URL 4.23 Internet 2 <http://www.Internet2.edu/>

REFERENCES

Baumann, J., 1999, Geotechnology spices up France's premier yacht race. Geofeature: GIS/Web-based Mapping. *GEOEurope*, **8**, (4), pp. 38-39.

Crampton, J., 1995, Cartography resources on the World Wide Web. *Cartographic Perspectives*, (22), pp. 3-11.

Crampton, J., 1998, The convergence of spatial technologies. *Cartographic Perspectives*, (30), pp. 3-5.

CyberAtlas, 1999a, The Big Picture, demographics: Worldwide Internet users to pass 500 million next century <http://cyberatlas.Internet.com/big_picture/demographics/article/0,1323,5911_200001,00.html> (accessed 30.11.1999).

CyberAtlas, 1999b, The Big Picture, geographics: The World's online populations <http://cyberatlas.Internet.com/big_picture/geographics/article/0,1323,5911_151 151,00.html> (accessed 24.11.1999).

Cyber Dialogue, Inc., 1999, American Internet User Survey, July 1999 <http://www.cyberdialogue.com/free_data/index.html> (accessed 12.10.1999).

Elzakker, C.P.J.M. van & Koussoulakou, A., 1997, Maps and their use on the Internet. In *Proceedings, Volume 2, of the 18th International Cartographic Conference ICC97, Stockholm*, edited by Ottoson, L. (Gävle: Swedish Cartographic Society), pp. 620-627.

Gebb, L., 1999, Personal communication, December 1999.

Hargittai, E., 1999, Weaving the Western Web: explaining differences in Internet connectivity among OECD countries. *Telecommunications Policy*, **23**, (10/11) (See also <http://www.princeton.edu/~eszter/hargittai-westernweb.pdf> (accessed 14.10.1999)).

Kehoe, C., Pitkow, J., Sutton, K., Aggarwal, G. and Rogers, J.D., 1999, Results of GVU's Tenth World Wide Web User Survey. *Graphics Visualization and Usability Center* <http://www.cc.gatech.edu/gvu/user_surveys/survey-1998-10/tenthreport.html> (accessed 12.10.1999).

Lambrecht, C. and Tzschaschel, S., 1999, National Atlas of the Federal Republic of Germany. In *Touch the Past, Visualize the Future. Proceedings 19th International Cartographic Conference, Ottawa, Canada*, edited by Keller, C.P., Section 3: Atlases (Ottawa: Organizing Committee for Ottawa ICA 1999) CD-ROM.

McGranaghan, M., 1999, The Web, cartography and trust. *Cartographic Perspectives*, (32), pp. 3-5.

Peterson, M.P., 1997, Cartography and the Internet: introduction and research agenda. *Cartographic Perspectives*, (26), pp. 3-12.

Peterson, M.P., 1999, Trends in Internet map use A second look. In *Touch the Past, Visualize the Future. Proceedings 19th International Cartographic Conference, Ottawa, Canada*, edited by Keller, C.P., Section 5: Capitalizing on new technologies (Ottawa: Organizing Committee for Ottawa ICA 1999) CD-ROM.

Press, L., Foster, W.A. and Goodman, S., 1999, Surveying the global diffusion of the Internet. Paper presented at the *INET'99 Conference*, 22 25th June 1999, San José, California <http://www.isoc.org/inet99/proceedings/posters/125/index.htm> (accessed 12.10.1999).

Sollberger, A., 1999, Vom Kartographieverlag zum Medienunternehmen am Beispiel der Kartographischen Online Plattform (KOP). In *Proceedings Symposium Web.mapping.99*, Karlsruhe 18.11.1999, pp. XI.1-XI.11.

Stähler, P., 1999, Potentiale des vernetzten Medienmanagements. In *Proceedings Symposium Web.mapping.99*, Karlsruhe 18.11.1999, (Karlsruhe: Fachhochschule), pp. X.1-X.14.

Thoen, B., 1999, Internet turns traditional economic models upside down. *GeoWorld*, April 1999, pp. 26-27.

UNDP, 1999, Globalization with a human face. Human Development Report 1999 <http://www.undp.org/hdro/report.html> (accessed 24.11.1999).

CHAPTER FIVE

Cartographic principles

Menno-Jan Kraak

5.1 VISUALISING GEOSPATIAL DATA

Maps are there to answer questions. They should offer solutions to questions like "Where can I find…?", "How do I get to…?", "What feature can be found at…?", or "Where else do I find that feature?" Maps have to be well designed to be able indeed to answer questions like those above. If the translation from data to graphics is successful the resulting maps are the most efficient and effective means of transferring geospatial information. The map user can locate geographic objects, while the shape and colour of signs and symbols representing the objects inform him about their characteristics. They reveal spatial relations and patterns, and offer the user insight in and an overview of the distribution of particular phenomena. Many illustrations in this book demonstrate an important quality of maps: their ability to offer an abstraction of reality. They simplify by selection, but at the same time they put, when well designed, the remaining information in a clear perspective. The interactive web environment adds another dimension to making maps, which can be experienced at the accompanying website (*URL 5.1*). Designing maps for this environment is treated in Chapter 7.

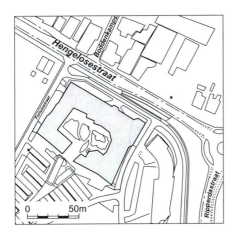

Figure 5.1 The map (left) versus the aerial photograph (right).

The typical map characteristics are well illustrated when one puts the map next to an aerial photograph or satellite image of the same area. Products like these convey all the information captured by the devices used. Figure 5.1 shows an aerial photograph of the ITC building in Enschede, the Netherlands and a map of the same area. The photograph shows all visible objects, including parked cars, a small temporary building, etc. From the photograph it becomes clear that the weather has influenced its contents: the shadow to the north of the buildings obscures other information. The map only gives the outlines of buildings and the streets in the surroundings. It is easier to interpret because of selection and classification of the data, and the symbolisation that highlights specific objects. Additional information, not available in the photograph, has been added, such as the name of a major street: "Hengelosestraat". Other non-visible data, like cadastral boundaries or sub-surface features like sewage pipes, could have been added in the same way. However, the map also demonstrates that selection means interpretation, a subjective activity. In certain circumstances a combination of photographs and map elements can be very useful.

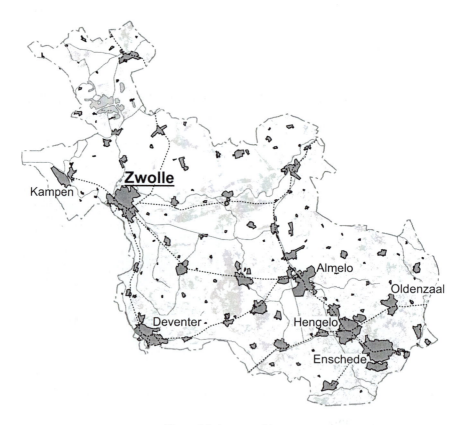

Figure 5.2. A topographic map.

The effectiveness of a map depends, next to the design, on its scale. For certain applications one will need very detailed maps (large scale maps – such as the one in Figure 5.1; 1 cm is 0.035km) while others only require overview maps (small scale maps, of which Figure 5.2 is an example; 1 cm is 8km). The map scale is the ratio between the distance on a map and the corresponding distance in reality. Scale indications on maps can be given in words like one-inch-to-the-mile, or in a representative fraction like 1:50.000 (1 cm on the map = 50.000 cm (or 500m) in reality), or by a graphic representation like a scale bar as given in Figure 5.1.

Traditionally maps are divided into topographic and thematic maps. Topographic maps visualise, limited by scale, the Earth's surface as accurately as possible. This will include for instance infrastructure (e.g. railroads and roads), land use (e.g. vegetation and built-up area), relief, hydrology, geographic names and a reference grid. Figure 5.2 shows a topographic map of the Dutch province of Overijssel.

Figure 5.3 A socio-economic thematic map: the population of Overijssel.

Thematic maps as shown in Figures 5.3 and 5.4 represent the distribution of particular themes. Figure 5.3 shows a population map of Overijssel and Figure 5.4 a map of the province's drainage areas. The first is an example of a socio-economic map and the second an example of a physical map. As can be noted both thematic maps also contain information found in the topographic map, since in order to be able to understand the theme represented one should be able to locate it as well. The amount and nature of topographic information required depend on the map theme. In general a physical map will need more topographic data then most socio-economic maps which normally only need administrative boundaries. The map with drainage areas needs the rivers and canals, while relief information is also normally added. Today's digital environment has diminished the distinction between topographic and thematic maps. Often both topographic and thematic maps are stored in the databases as layers. Each layer contains data on a particular topic, and the user is able to switch layers on or off at will.

Figure 5.4 A physical thematic map: the drainage areas of Overijssel.

The design of topographic maps is mostly based on conventions, of which some date back to the nineteenth century. Examples are water in blue, forests in green, major roads in red, urban areas in black, etc. The design of thematic maps, however, is based on a set of cartographic rules, also called cartographic grammar (Dent, 1996; Robinson *et al.*, 1995; Kraak and Ormeling, 1996). Nowadays maps are often products of a GIS. GIS is about data integration, spatial analysis and models. If one wants to use a GIS to solve a particular geo-problem this often involves the combination and integration of many different data sets. For instance if one wants to quantify land use changes the combination of two data sets, each from a different period, can be done via an overlay operation. The result of this spatial analysis could be a map showing the difference between both data sets.

The parameters used during the operation are based on models developed by the application at hand. It is easy to imagine that maps can play a role during each of these steps while working with a GIS. From this perspective maps are no longer the final product they used to be. They can be created just to see which data are available in the spatial database, or to show intermediate results during a spatial analysis, and of course to present the final outcome. The whole process of dealing with data is also known as the spatial data handling process. Spatial data handling stands for the acquisition, storage, manipulation and visualisation of spatial data in the context of particular applications.

In Figure 2.1 *"How do I say what to whom, and is it effective"* was introduced. This guides the cartographic visualisation process, and as such summarises what is called the cartographic communication process. Especially when dealing with maps that are created in the realm of presentation cartography it is important that they adhere to the cartographic design rules. This is to guarantee that they are easyly and well understood by the map users. How does this communication process work? Figure 5.5 demonstrates this process and the following section will elaborate some aspects in more detail. It starts with information to be mapped (the *What* from the sentence).

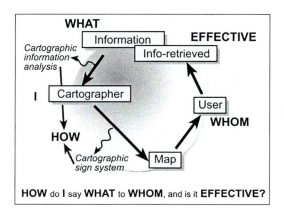

Figure 5.5 The cartographic communication process.

5.2 CARTOGRAPHIC INFORMATION ANALYSIS

Before anything can be done the cartographer should get a feel for the nature of the information, since this determines the graphic options. This is done via cartographic information analysis. Based on this knowledge the cartographer can choose the correct symbols to represent the information in the map. The cartographer has a whole toolbox of visual variables available to match symbols to the nature of the data. The French cartographer Bertin (1967) developed the basic concept of the theory leading to a good map. He did this with his publication "Sémiologie Graphique". The rules described by Bertin should be interpreted as guidelines for map design. If you would give ten professional cartographers the same mapping task, and each would apply the rules it could still result in ten different maps. For instance, if the guidelines indicate to use colour it is not said which colour. All ten maps could be of good quality. Returning to the scheme, the map (the *How* in the sentence) is read by the map user (the *Whom* in the sentence) who extracts a certain amount of information from the map, represented by the box entitled "Information Retrieved". From the figure it becomes clear that the boxes "Information" and "Information Retrieved" do not exactly coincide. This means the information derived by the map user is not the same as the information the cartographic communication process started with. This has several causes. It is possible the original information is indeed partly lost and/or additional information has been added during the process. Loss of information could be deliberately caused by the cartographer, with the aim of emphasising the remaining information. Another possibility is that the map user was unable to understand the map fully. Information gained during the communication process could be due to the cartographer, who added extra information to strengthen the already available information. It is also possible that the map user has some prior knowledge on the topic or area, which allows him or her to combine this prior knowledge with the knowledge retrieved from the map.

To find the proper symbology for a map one has to execute a cartographic data analysis. The core of this analysis process is to access the characteristics of the data to find out how they can be visualised. The first step in the analysis process is to find a common denominator for all the data. This common denominator will then be used as the title of the map. For instance if all data are related to hydrology the title will be "Hydrology of...". Next the individual components, such as those that relate to watersheds, should be accessed and their nature described. Later, these components should be found in the map legend. Analysis of the components is done by determining their so-called measurement scale. Data will be of a qualitative or quantitative nature. The first type of data is also called nominal data. Examples are the difference between the languages spoken (e.g. English, Swahili, Dutch...), the differences between soil types (e.g. sand, clay, peat....) or the difference between land use categories (e.g. arable land, pasture...). The difference between the data types is based on qualities only. In the map qualitative data is classified according to the methods in use in a particular discipline, for example a soil classification system. Basic geographic units will be more or less homogeneous areas that adhere to one of the soil types recognised by the soil classification. In between qualitative and quantitative data one can distinguish ordered data. These data are measured along an ordinal scale, and are based on hierarchies. For instance one knows that one value is more or less then another, such as cool versus warm. Another example is the differences among roads: highway, main road, secondary road and track. Quantitative data can be measured along an interval or ratio scale. For data measured on an interval scale the exact distance between values is known, the zero is arbitrarily chosen and negative values are possible. Temperature is an example: 40°C is not twice as warm as 20°C. For data fitting the ratio scale an absolute zero is known, e.g. temperature measured in degrees Kelvin. Another example is income: someone earning $100 has twice as much as someone with $50. In maps quantitative data are often grouped together according to a mathematical method.

- Nominal   Data on different nature/identity of things
- Ordinal   Data with a clear order, though not quantitatively determined
- Interval   Quantitative information with arbitrary zero
- Ratio   Quantitative data with absolute zero

Let us consider why the maps in Figures 5.3 and 5.4 look like they do. Table 5.1 holds data to create a map of Overijssel's watersheds. It lists the separate drainage areas and it also defines the common theme of the map: the provincial drainage areas. This could be the title of the map. The areas are considered to be the geographic component. As such, their differences can be measured on a nominal scale, as in a political map in an atlas. It is important to know how many elements the geographic component has. The table shows 14 areas. If there are too many (in relation to the scale) the map could become over-burdened with boundaries. Imagine a table with over 2500 areas. It could be problematic to show them all on a map on a page in this book. In that case one should consider grouping the areas together at a higher aggregation level.

Table 5.1 The data behind the map in Figure 5.4: provincial drainage areas.

Id	Name	Id	Name
8	Bacaem van Vollenhove	6	Weteringen
7	Reest-Meppelerdiep	11	Schipbeek
3	Ommerkanaal	13	Friesland
2	Afwateringskanaal	14	Randmeren
5	Vecht	12	IJssel
1	Dinkel	9	Zwartewater
4	Regge	10	Twentekanaal
		15	Unknown

Another simple example is given in Table 5.2. It shows the number of inhabitants of Overijssel's municipalities in 1998. This could also function as the future map title. The table shows two components, again the geographic component (the names and IDs of the 45 municipalities) and the number of inhabitants per municipality. This component is of a quantitative nature (measurement scale is ratio because the zero is absolute). It shows absolute numbers that range from 2744 inhabitants in Diepenheim to 148161 inhabitants for Enschede. It is important to know the range of the quantities to be represented in the map, because this has implications for the symbolism to be chosen. The next section will explain this in more detail.

5.3 THE CARTOGRAPHIC SIGN SYSTEM

A map image consists of point symbols, line symbols, area symbols and text. The symbols can vary in their appearance as can be seen in Figure 5.6. The figure shows symbols in different sizes, shapes and colour. The points can represent individual objects such as the location of shops or could refer to values that are representative for an administrative area. Lines can vary for instance in colour to show the difference between a boundary and a river, or in form to show the difference between railroads and roads. Areas follow the same principles: for instance difference in colour to distinguish between different vegetation stands. Although the changes are only limited by fantasy they can be grouped together in a few categories.

Table 5.2 The data behind the map in Figure 5.3: provincial municipal population.

Id	Name	Inhabitants	Id	Name	Inhabitants
141	Almelo	65632	165	Holten	8761
143	Avereest	14971	195	IJsselham	5529
144	Bathmen	5241	191	IJsselmuiden	14536
146	Borne	22051	166	Kampen	32161
194	Brederwiede	12296	168	Losser	22722
148	Dalfsen	17376	169	Markelo	7135
142	Delden, Ambt	5444	170	Nieuwleusen	8270
179	Delden, Stad	7344	173	Oldenzaal	30670
159	Den Ham	14847	174	Olst	9311
149	Denekamp	12300	175	Ommen	16493
150	Deventer	70655	176	Ootmarsum	4472
151	Diepenheim	2744	177	Raalte	28281
152	Diepenveen	10530	178	Rijssen	26083
153	Enschede	148360	180	Staphorst	15178
154	Genemuiden	8747	181	Steenwijk	22396
156	Goor	12338	183	Tubbergen	19793
157	Gramsbergen	6413	186	Vriezenveen	19744
158	Haaksbergen	23706	188	Weerselo	9305
160	Hardenberg	34891	189	Wierden	22946
161	Hasselt	7525	190	Wijhe	7494
162	Heino	7888	192	Zwartsluis	4466
163	Hellendoorn	35544	193	Zwolle	102622
164	Hengelo	78306			

Bertin (1967) distinguished six categories, which he called the visual variables. They are size, value, texture (grain), colour, orientation and form. These visual variables can be used to make one symbol different from another one. In doing this, map makers in principle have a free choice. They do not have that choice when deciding where to locate the symbol in the map. The symbol should be located where features belong. The grouping in these categories is based on how they stimulate a certain perceptual behaviour with the map user. The behaviours are quantitative, selective, ordered and associative. Confronted with differences in (grey) value one will experience differences in order, and (perhaps) quantity. Maps of for instance population density use this principle. Form and colour allow one to differentiate among qualitative data, because they give an impression of differences among things, but not differences in relative importance. In Figure 5.4 pattern is used to differentiate among the drainage areas of Overijssel. Size is a good variable to use to show amounts (ratio data), as can be seen in Figure 5.3, while value functions well in mapping data measured according to the interval scale. The scheme in Figure 5.7 shows the relation between the individual graphical variables and their perceptual properties.

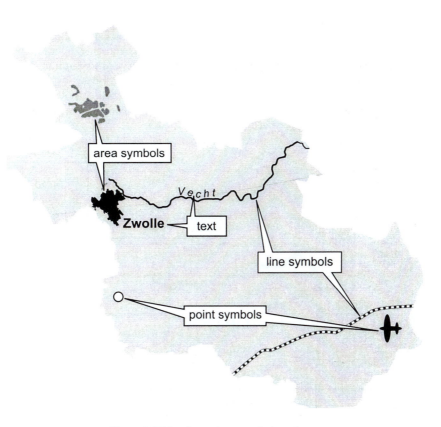

Figure 5.6 Point, line and area symbols, and text.

Now we are aware of two major elements of the visualisation process, the analysis of the data to be mapped and the characteristics of the visual variables. The next step is to combine the results of the analysis with the perception properties of these variables to finally create the maps. The data in Table 5.1 are qualitative. In the final map the user should perceive these qualities. This means we have to find a graphic variable that has associative perception properties. Looking at the scheme in Figure 5.7 the variables colour, shape and orientation have these properties. Colour is the best choice, since it also has selective perception properties. It allows one easily to see what belongs together and still allows overview. For Table 5.2, representing the number of people of each municipality in Overijssel, it was concluded that the data had a quantitative nature, with absolute numbers (ratio scale). A graphic variable with a quantitative perception property is needed. From the scheme in Figure 5.7 it follows that only the graphic variable size is suitable for this purpose.

The maps in Figures 5.3 and 5.4 represent very common types of thematic map. The second is known as a chorochromatic or mosaic map representing nominal data, such as soils, geology, land use. A political map of Africa showing all countries in a different colour also belongs to this category. The proportional point symbol map is another map often used. This map can represent both qualitative and quantitative data. The symbol size represents quantities, while its colour represents qualities. Other common map types are the choropleth map and the isoline map. The first is basically a density map representing ordinal, interval or relative quantitative data such as inhabitants per unit area or per capita income per administrative unit. Contour maps and climate maps are probably the most well known representatives of the second type. Each line in an isoline map has the same value, e.g. in height or temperature.

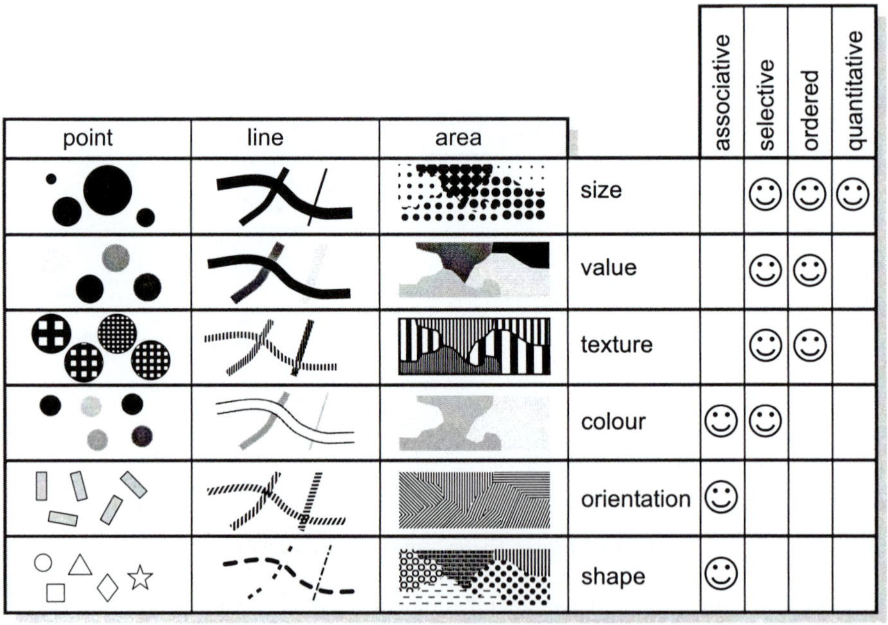

point	line	area		associative	selective	ordered	quantitative
			size		☺	☺	☺
			value		☺	☺	
			texture		☺	☺	
			colour	☺	☺		
			orientation	☺			
			shape	☺			

Figure 5.7 The graphic variables.

Why does it make sense to follow the guidelines explained above? This is best illustrated by some examples showing good and bad applications of the cartographic design theory. Let us revisit the watershed map in Figure 5.4. Figure 5.8a shows it again. Fifteen categories are distinguished, each represented by a different pattern in the figure in this book (colour on the website and in Plate 1). The map in 5.8b seems to follow the same principle, yet the map user will get a different message from this map. While the map in 5.8a pays more or less equal attention to all watershed areas, the map in 5.8b does not. A clear emphasis is given to the dark area in the centre of the map. The choice of one dark pattern and all the rest as light patterns (on the website and in Plate 1 one bright colour and all the rest pastel colours) will divert attention to that area. Of course, the map shown in 5.8b can be suitable if one really wants to highlight that particular area. Figure 5.8c shows the same data, but now only mapped with the visual variable value. Again, this is inappropriate. The result is a choropleth map, which offers the user the impression of order. Darker areas are seen as more important than lighter areas.

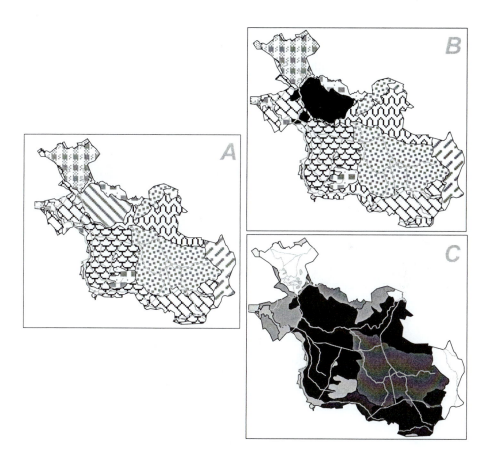

Figure 5.8 Right and wrong, mapping qualities: a) the correct map; b) incorrect application of the graphic variable; c) incorrect graphic variable.

Another example to show why it makes sense to apply the cartographic theory is presented by Figure 5.9. Here the same quantitative data are displayed in three maps. The first (5.9a) gives the correct representation of the data, population density for Overijssel's municipalities. Value has been used to display the density from low (light tints) to high (dark tints). The map reader will automatically and at a glance associate the dark colours with high density and the light values with low density. In Figure 5.9b value has been used but the tints have not been ordered in sequence. The first impression of the map reader would be to think that the dark areas represent the areas with the highest density. A closer look at the legend shows that this is not the case. Figure 5.9c shows the effect of the application of an incorrect graphical variable. Different patterns (colours on the website) have been used. The reader will have to go to quite some trouble to find out where in the province the highest population density can be found. If one really studies the legends of the maps described as wrong in both 5.8 and 5.9 the information can be derived but it would take some effort and lots of time to do so. The proper application of the cartographic guidelines will guarantee this will go much more smoothly (e.g. faster and with less chance of mistakes).

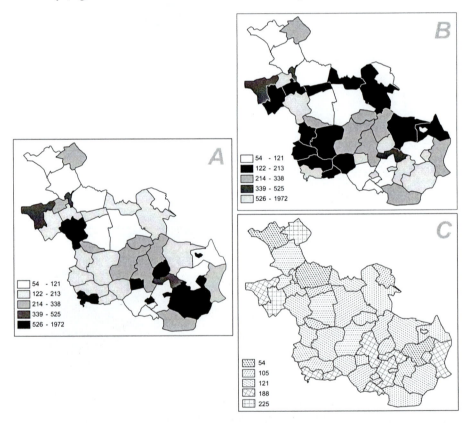

Figure 5.9 Right and wrong, mapping quantities: a) the correct map; b) incorrect application of the graphic variable; c) incorrect graphic variable.

The maps like those in Figures 5.8a and 5.9a are correct from a cartographic grammar perspective. However, the maps in most figures in this chapter lack the extra information needed to fully understand them. Each map should have, next to the map image, a title, informing the user about the topic visualised. A legend is necessary to understand how the topic is represented. Additional marginal information to be found on a map should be a scale indicator, a north arrow for orientation, the map projection used, and some bibliographic data. The bibliographic data should give the user an idea when the map was created, how old the data used are, who has created the map and even what tools were used. All this information allows the user to get an impression of the total value of the map. This information is comparable with metadata describing the contents of databases. Figure 5.10 summarises these map elements. On paper maps these elements have to be shown. Maps presented on screen (and the WWW) often go without marginal information, partly because of space constraints. However, on-screen maps are often interactive, and clicking on an object will reveal information from the database. Legends and title are often available on demand.

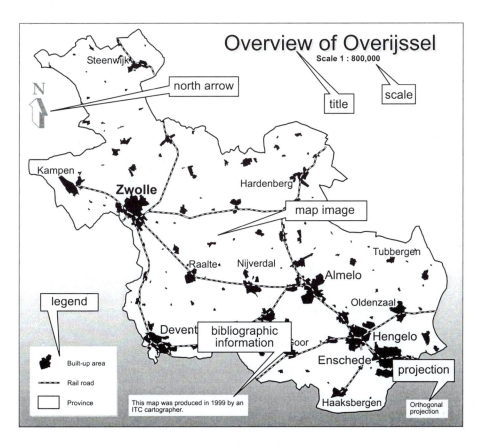

Figure 5.10 The map and its marginal information.

Maps constructed using the basic cartographic guidelines are not necessarily appealing maps. Although well constructed, they could still look sterile. The design aspects required to create appealing maps also have to be included in the visualisation process. Appealing in a communicative sense does not only mean having nice colours. One of the keywords here is contrast. Contrast will increase the communicative role of the map since it will create a kind of hierarchy in the map contents, assuming that not all information is of equal importance. This design trick is also known as visual hierarchy or the figure-ground relation. This is best understood when looking at both a well designed and a poorly designed map as shown in Figure 5.11. The purpose of this map is to indicate the location of the municipality of Enschede in Overijssel. The left map just shows boundaries with a name. The second map first sets Overijssel free from the background, while Enschede itself looks elevated above Overijssel. The map user's eyes will certainly be attracted to Enschede's location in the province. There are many, more complex ways in which maps can be made effective and appealing. Some of them are described in Chapter 7 of this book and further illustrated in Chapters 8 to 12 by examples from existing websites.

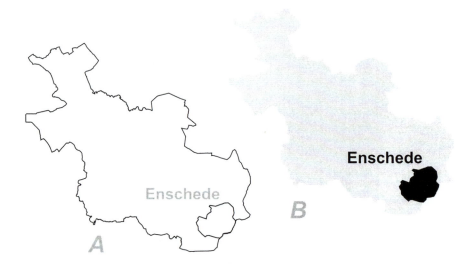

Figure 5.11 Appealing maps, the location of the municipality of Enschede in the province of Overijssel: a) a map without contrast (visual hierarchy); b) a map with contrast.

5.4 MULTI-DIMENSIONAL DISPLAY

The previous section dealt with the basics of cartography. However, the more technology evolves the better it became possible to capture and model the real world. This world is three-dimensional. It is obvious that one needs three-dimensional visualisations as well. To gain an insight in the more complex data relations in the geospatial databases one cannot do without three-dimensional visualisations. The Web offers an environment to create and distribute three-dimensional graphics rather easily.

In the past cartographers have always struggled with the third dimension since it had to be mapped on a two-dimensional sheet of paper or a flat screen. Especially if one studies the history of relief mapping these problems become clear. Shaded relief maps, based on for instance contour lines can be very convincing in portraying the landscape. However, when converted into a three-dimensional space the 3D perception can be increased. Looking at such a representation as given in the perspective view in Figure 5.12 one can immediately imagine that this will not always be effective. Certain objects in the map will easily disappear behind other objects. Interactive functions to manipulate the map in the three-dimensional space to look behind some objects are required. These manipulations include panning, zooming, rotating and scaling. Scaling particularly along the z-axis of the model is needed, since if the area covered by the model is fairly large, the vertical height differences are very small compared to the horizontal distances. On a model of the Himalayas with true vertical scale even Mt. Everest would not be much more impressive than a Dutch hill! With today's software it is relatively easy to change vertical scale and to see or calculate relations between the terrain model and other data available.

Figure 5.12 Relief: left an orthogonal view and right a perspective view.

Thematic data can also be viewed in three dimensions. This can result in dramatic images, which will be long remembered by the map user. Figure 5.13 shows the population data of Overijssel as displayed in Figure 5.3 in three dimensions. Instead of a symbol that depicts the number of people living in a municipality the height of this municipality is used to display amounts. Since data in two-dimensional maps is often classified in a few categories the relations among the geographic objects are better seen. The image clearly shows that both Enschede (the column in the lower right) and Zwolle, the province's capital (the column in the upper left) are the province's two largest cities. Manipulating this model in three-dimensional space will increase this insight even better. On the Web the VRML-environment offers opportunities to manipulate these three-dimensional visualisations, and even include different levels of detail and hyperlinks (see Appendix A).

Advances in geospatial data handling are not just related to the third dimension. Dealing with time has also become part of the daily routine now. This is due to the increase in the availability of data captured at different periods in time. Next to this data abundance, the GIS community wants to deal with real world processes as a whole, and no longer with single time-slices. To visualise models or planning operations is no longer efficient with static maps only. From a visualisation perspective, animations can be the solution. Based on the characteristics of geospatial data animations can depict changes in space (position), in place (attribute), or in time. Cartographic animations will help to improve the understanding of processes behind the changes shown.

Figure 5.13 Overijssel in a VRML environment.

It is possible to distinguish between temporal and non-temporal animations. The first show changes of spatial patterns in time. Most familiar examples are seen when looking at a weather broadcast: animations with moving clouds or changing temperatures (see Chapter 11). Other examples are those of the changing Dutch coastline from Roman times until today or the changing municipal boundaries of Overijssel. Figure 5.14 shows some animation frames. Non-temporal animations offer the viewer a different look at a particular data set. In these animations location, attribute and time are fixed. They show the same data, but from a different graphic perspective. An example is a simulated flight through a landscape. The VRML-environment on the Web discussed above offers opportunities to walk or fly through the model on display. The web-site version of figure 5.13 allows the reader to practise this.

For the user of a cartographic animation, it is important, similar to three-dimensional environments, to have tools available that allow for interaction while viewing the animation. Just to see the animation play will often leave users with many questions about what they have seen. Only a replay is not sufficient to answer questions like "What was the position of the coastline in the north during the 15th century?" Most general software to view animations already offers facilities such as "pause", to look at a particular frame, and "(fast-) forward" and "(fast-) backward", to go to a particular frame. More options have to be added, such as a possibility to directly go to a certain frame based on a task like: "Go to 1850". Web versions of most media players are more limited in their interaction options, although it is likely this will improve quickly. Just as any other map, animated maps need a legend to explain the meaning of the map symbols used. However, the legend can have a dual function now. Besides being a tool for explanation it can be a tool for navigation, by moving a slide bar to go to the period of interest. This is an example of the "control-panel legend" discussed further in Chapter 7.

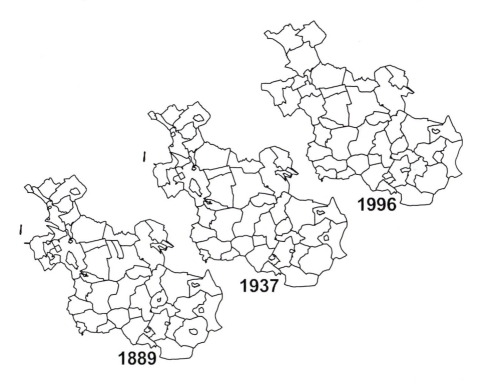

Figure 5.14 Animation frames: change of municipal boundaries.

CHAPTER SIX

Publishing maps on the Web

Barend Köbben

6.1 INTRODUCTION

The World Wide Web is a relatively new means of publishing maps and although
nowadays millions of people use it every day, not many of those users know how
it actually works. Many of them, quite sensibly, do not really care. But for the
readers of this book some basic knowledge about the technical background of the
Web can be helpful in understanding the different mapping possibilities available.

In this chapter, a short overview is given of how the World Wide Web
sprouted from the Internet and of the file formats and communication protocols
that are its building blocks. It will become clear how the basic set-up allows the
publication only of static maps with very limited interactivity and how the
possibilities can be broadened by adding server-side applications or client-side
plugins and applications such as Java applets or scripts. ESRI's ArcView Internet
Map Server will be used as an example of how all these techniques can be
combined to offer mapping and GIS functionality on the Web.

6.2 INTERNET AND THE WORLD WIDE WEB

To understand the World Wide Web, one needs to understand Internet, because the
Web is a subset of the broader concept of the Internet. The Internet has its roots in
the very modest first start of computer networking called ARPAnet. This was a
network created for the Advanced Research Projects Agency and funded by the
U.S. Department of Defense. Its purpose was twofold: first to create the possibility
of scientists residing in different locations to work together and share information
and secondly, to spread vital information over a range of computers in different
locations in order to diminish its vulnerability to nuclear attacks. The network was
built as a distributed, decentralised system with each node of equal importance and
the computers in the network communicated with each other using a so-called
"communication protocol" (called NCP – Network Control Protocol), which
defined how the systems should request, send and retrieve information from each
other. By the end of 1969, this grandmother of the Internet connected four host
computers at the University of California Los Angeles, the University of California
Santa Barbara, the University of Utah and Stanford Research Institute (Salus,
1995). As a nice gesture to the history of web cartography, this configuration was
mapped at the time, as can be seen in Figure 6.1.

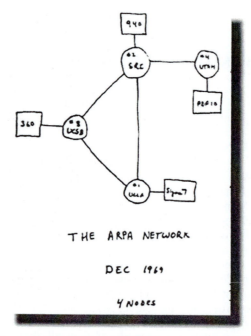

Figure 6.1 Sketch of the initial ARPAnet (from Salus, 1995).

Gradually, the ARPAnet grew to encompass many North American computers and in 1983 it switched to a new communication protocol called TCP/IP (for Transmission Control Protocol/Internet Protocol). Around the same time, USENET, the Unix users' network that carried most of the e-mail and news of that time and the European side (EUnet) were also converted to use TCP/IP. All kinds of other communication protocols, such as TelNet (for terminal services), FTP (for file transfers) and SMPT (for e-mail communication) were redirected to run "on top of" TCP/IP. This is considered to be the birth of the Internet as we know it (*URL 6.1*).

The reason TCP/IP is so important today is that it allows stand-alone networks to be connected to the Internet or linked together to create private intranets. Devices called routers or IP routers physically connect the networks that comprise an intranet. A router is a computer that transfers packets of data from one network to another. On a TCP/IP intranet, information travels in discrete units called IP packets or IP datagrams. TCP/IP software makes each computer attached to the network a sibling to all the others; in essence, it hides the routers and underlying network architectures and makes everything seem like one big network. Connections to the Internet are identified by 32-bit IP addresses, organised as dotted decimal numbers, for example 192.92.92.70. Given a remote computer's IP address, a computer on the Internet can send data to the remote computer as if the two were part of the same physical network (for more details, see *URL 6.2*). To make the addressing system more versatile, a system of Domain Name Servers (DNS) has been established. These map the numerical IP addresses to a hierarchical system of logical names. The IP address 192.92.92.70 thus is mapped to the IP name "kartoweb.itc.nl", indicating a computer called "kartoweb" within the network of the ITC, which in turn is part of the Dutch (nl) section of the Internet. The DNS system is in fact a hierarchical distributed database, with each level administering its own subsection. As shown in Figure 6.2, higher level administrators can use *DNS aliasing* to re-route requests to lower level domains, so that the computer with the IP number 192.92.92.70 answers to the name "www.kartografie.nl" (a website maintained by the Netherlands Cartographic Society) as well as to "kartoweb.itc.nl".

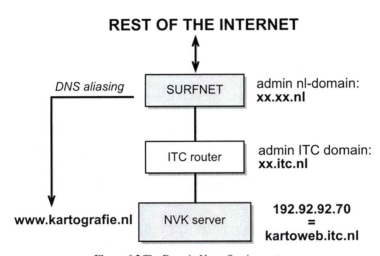

Figure 6.2 The Domain Name Serving system.

Somewhat confusingly, besides Internet there is this thing called the World Wide Web (WWW or simply "the Web"). It was originally started by Tim Berners-Lee at CERN (the European Laboratory for Particle Physics) to enable the dissemination through the Internet of information among the various research facilities of that institute. In order for this to be possible the communication protocols of the Internet were not sufficient, what was also needed were standardised ways to store the information. For this purpose, the HyperText Mark-up Language (HTML – see Appendix A.2) was devised. A prototype of HTML and a new communication protocol HTTP (HyperText Transfer Protocol), specifically geared towards finding and retrieving HTML and other documents from various servers (to be run "on top of" TCP/IP), was agreed upon in 1994 (*URL 6.3*).

In practice, the Web is now a vast collection of interconnected documents, spanning the world, which merges the techniques of networked information and hypertext to make a powerful global information system. It makes information accessible over the network as part of a seamless hypertext information space (Peterson, 1996).

6.3 BASIC PUBLISHING ON THE WEB

One of the strong points of the Web is the fact that it is virtually platform-independent. The data can be served by computers running AppleOS, Unix, Linux, BeOS, VMS or Windows, to name but a few. For retrieving and viewing the information, there are so-called *web browsers* available at little or no cost for every type of computer in existence, including even game computers, organisers and mobile phones. The underlying TCP/IP *communication protocol* makes sure all these computers can interact. But in order for all the browsers to be able to interpret and show the content correctly, there also are a limited number of standardised *data formats* that have to be used to store the information.

6.3.1 Basic web data formats

The main Web format is the HTML document mentioned earlier. In any hypertext document, if you want more information about a particular subject mentioned, you can usually just "click" on it to read further details. In fact, documents can be and often are linked to other documents by completely different authors, stored on computers in totally different locations. The HTML format takes this a step further because it supports hyper*media*, a superset of hypertext – it is any medium with pointers to other media (*URL 6.4*). This means that browsers not only display and link textual information, but also might be used to view images and animations and to listen to sounds.

HTML is a non-proprietary format and can be created and processed by a wide range of tools, from simple plain text editors – you type it in from scratch – to sophisticated WYSIWYG authoring tools. Converting information to HTML format is possible nowadays in most "office" environments (this term indicates much-used software for common office tasks, such as word processing, presentation and spreadsheets).

HTML is essentially plain text interspersed with so-called "tags" to structure text into headings, paragraphs, lists, hypertext links, etc. These tags are used by the browser that is used to interpret the HTML files to structure and layout the text. For example the HTML code:

```
<P>Hello <B>World</B><HR>How are <I>you</I>?<P>
<A HREF= "otherfile.html">Click here</A> to go.
```
would be shown as:

Hello **World**
How are *you?*
Click here to go.

Clicking on the words "Click here" would result in another file (stored using the name "otherfile.html") being loaded by the browser, interpreted and displayed on the screen.

To show graphics, so-called inline images can be embedded on-screen by referring to their files in the HTML code. For this, two raster formats are generally used: GIF (Graphics Interchange Format) and JPEG (Joint Photographics Expert Group). Both are raster formats, the former better suited for graphics having areas of solid colours, such as graphs and maps, and the latter more appropriate for half-tone graphics, such as photographs (for a more detailed description, see Appendix A.3).

6.3.2 Basic web publishing set-up

To publish maps and information on the Web, one needs only very few resources. The actual requirements depend largely on the flexibility and sophistication needed, but the minimum set-up includes the following elements:

1. A so-called "backbone" connection to the Internet. This is a fixed and high capacity connection. The data transfer capacity of the more usual modem connections is insufficient. Such backbones can be found at major companies, educational and government institutes, etc. When no such server is at hand, one can hire facilities from commercial Internet Service Providers.
2. One or more computers using the Internet connection mentioned in 1.
3. An application that understands the HTTP protocol and listens for requests coming in for documents stored on the servers connected to it. It basically waits for HTTP "GET" commands and will respond by transferring the files asked for to the requesting network node by using the HTTP "PUT"-command. This application is called a *web server*.
4. Basic text editing tools or a WYSIWYG editor to author the HTML files.
5. Basic graphic tools to make the graphics. These graphics should be specially designed for web use (see Chapter 7).

The last point is especially important, because of what currently is the major disadvantage of using the Web: the limited transfer capacity of the Internet. The hardware infrastructure is at present not sufficient for the large number of users, resulting in poor data transfer rates, especially when using a modem connection as the majority of users do. This is quickly becoming a genuine problem, such that the abbreviation WWW is sometimes said to mean "World Wide Wait". In order to offer users at least bearable response times, the files transferred over the Internet need to be as small as possible, making the use of large format, high-resolution graphic files undesirable.

The combination of the server and browser software enables the retrieval of static information only, that is the user can review text and graphics as and how they are stored on the server. If for example one wants to publish thematic maps derived from certain statistical data, one would have to prepare these maps in advance and include them as embedded images from which the user can choose. Thus, elements such as the layout, the visualisation techniques and classification would be fixed and the user cannot change these. This basic set-up is shown in Figure 6.3.

What happens in this basic set-up can be described as a simple series of requests and responses. It starts by the user providing the browser with the Internet address of the information he or she wants to view. This address will be in the form of a URL, a Universal Resource Locator, for example "http://kartoweb.itc.nl/index.html". This URL could be typed in by the user or could be the result of a hyperlink from some other document. It will result in the browser sending a request (using the http "GET" command) to the server located at the IP address kartoweb.itc.nl for the file called "index.html". The server at this location will respond to this request by sending the file (using the http "PUT"command) to the IP address of the browser sending the request. This browser will read the HTML file and will interpret the tags and commands inside that document to show the user the information in the file. When, as in the example given in Figure 6.3, the code for an embedded image (the <IMG>–tag) is encountered, an additional request is sent for the image file stated in the "SRC=" parameter of the tag. The image is received, displayed on the screen and the browser resumes interpreting the HTML code.

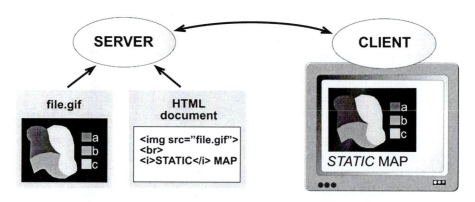

Figure 6.3 Basic web publishing.

Any image included in a HTML-file can be made to have certain "sensitive areas", defined by the coordinates of their bounding polygons. By clicking the mouse within such parts of the image the browser will link with another Internet location. Thus, a single image can provide different hyperlinks, each assigned to defined regions of the image. These images with "invisible buttons" are called *clickable images* or *image maps* (even though they are not always maps, but maybe logos or any other graphic). For an annotated example with more explanation, see the interactive version of Figure 6.4 on this book s website.

There are two kinds of clickable maps: "server-side" clickable maps and "client-side" clickable maps. With server-side maps, the coordinates of the bounding polygons are not stored in the presentation itself, but in a separate map file on the web server (hence the name). The disadvantages of this are that it creates extra data traffic between server and client and that one cannot test image maps locally, because a "live" connection to the supplier's server is needed. That is why recently the client-side clickable maps have evolved. Client-side image maps store the hyperlink information in the HTML document in which the image is embedded, not in a separate map file. When the user clicks a hotspot in the image, the associated hyperlink location is determined by the browser software and the user is transferred directly to that location. This makes client-side image maps faster than server-side image maps. Client-side image maps are nowadays supported by all major browsers.

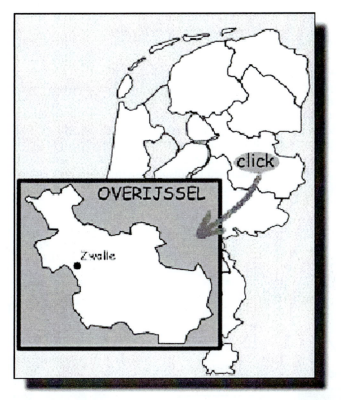

Figure 6.4 Screen dump of a clickable map (annotated interactive version on this book's website).

6.4 EXTENDING WEB MAP FUNCTIONALITY

The standard set-up described in the previous paragraph makes use of the configuration complying with the official standards of the World Wide Web, as maintained by its governing body, the World Wide Web Consortium (see *URL 6.5*). This set-up is quite limited, for example only the two raster formats mentioned (GIF and JPEG) could be used to show graphics. For mapping applications this severely limits the possibilities. Only static maps are possible, with no interactivity other than the clickable maps mentioned earlier. To overcome the limitations, many solutions are available. In the following paragraphs, additions to the basic system are described which could be realised at the client-side (i.e. the web browser) or at the server-side, or that could be a combination of both. One has to realise however, that many of the solutions discussed below require extensions to the standardised client-server functionality, thus undermining the platform-independence and making the information less generally useable.

6.4.1 Client-side functionality

Plugins

In the basic set-up described above, only three file formats are used: HTML for the text and layout and GIF or JPEG for the images. Using exactly the same client-server communication, other formats could theoretically be used, among them vector file formats. However, this requires that the web browser can interpret and display these file formats. For this purpose, most browser applications offer a mechanism that allows third-party programs to work together with the browser as a so-called *plugin*. This plugin technology is the simplest and probably the most used extension of the web functionality. Many thousands of plugins are available for almost as many file formats. The example in Figure 6.5 shows the use of the Acrobat plugin that Adobe devised for displaying their Portable Document Format (PDF), a popular document description format based on PostScript.

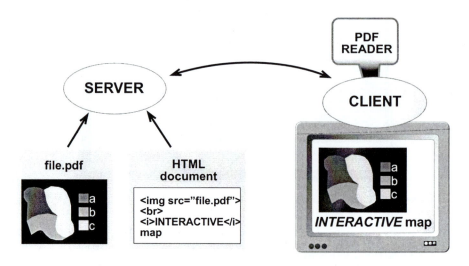

Figure 6.5 Plugin set-up (example of Adobe Acrobat PDF–reader).

As can be seen when comparing Figures 6.5 and 6.3, the only changes to the basic set-up are the addition of the plugin to the browser and the different extension of the filename in the "SRC=" parameter of the <IMG> tag. This extension allows the browser to recognise the file as not being a "native" format and to transfer the interpretation of the file to the appropriate plugin. The Acrobat PDF reader takes over control and allows the extended functionality of the PDF file: zooming and panning, printing at high resolution, selecting and querying text and graphics, etc.

The advantage of this system is the ready availability of plugins for very many file formats, thus making possible the distribution of almost every graphics file through the Web. Mostly these plugins are distributed free of charge by the proprietors of the file format.

The main disadvantage lies in downloading and installing these plugins and their all too frequent updates. While surfing the Web, the necessity to install a new plugin occurs almost as often as the need for a coffee. Furthermore, the plugins are programmed for specific operating systems and sometimes even for specific browsers only.

Java and JavaScript

The use of Java on the Web became widespread in a very short period. The main reasons for the rapid acceptance is the fact that it does not have the platform-dependence of the plugin solution and that the major players in the market are now including Java interpreters (so-called *virtual machines*) into their browser programs. Thus Web functionality can easily be extended by supplying small stand-alone Java programs or including Java scripts in the actual HTML code.

Java is an object-oriented programming language designed for fast execution and type safety (*URL 6.6*). Type safety means, for instance, that you cannot accidentally access memory (other than that reserved for the application).

Table 6.1: JavaScript compared to Java (from *URL 6.6*).

JavaScript	Java
Interpreted (not compiled) by client.	Compiled bytecodes downloaded from server, executed on client.
Object-based. No distinction between types of objects. Inheritance is through the prototype mechanism and properties and methods can be added to any object dynamically.	Object-oriented. Objects are divided into classes and instances with all inheritance through the class hierarchy. Classes and instances cannot have properties or methods added dynamically.
Code integrated with, and embedded in, HTML.	Applets distinct from HTML (accessed from HTML pages).
Variable data types not declared (loose typing).	Variable data types must be declared (strong typing).
Dynamic binding. Object references checked at runtime.	Static binding. Object references must exist at compile-time.
Cannot directly write to hard disk.	Can write to hard disk.

JavaScript and Java are similar in some ways but fundamentally different in others (see Table 6.1). Java's object-oriented model means that programs consist exclusively of classes and their methods. This makes Java programming more complex than JavaScript authoring. The JavaScript language resembles Java but does not have Java's static typing and strong type checking.

JavaScript supports most Java functionality plus extras such as predefined objects only relevant to running JavaScript in a browser. It is embedded directly in HTML pages and is interpreted by the browser completely at runtime. When the browser (or client) requests such a page, the server sends the full content of the document, including HTML and JavaScript statements, over the network to the client. The client reads the page from top to bottom, displaying the results of the HTML and executing JavaScript statements as it goes. Client-side JavaScript statements embedded in a HTML page can respond to user events such as mouse-clicks, form input, and page navigation. For example, you can write a JavaScript function to verify that users enter valid information into a form requesting a telephone number or zip code. Without any network transmission, the HTML page with embedded JavaScript can check the data entered and alert the user with a dialogue box if the input is invalid.

Using Java applets, the same code would be not incorporated in the HTML page, but would be pre-compiled into a so-called "Java class" and stored on the server. The HTML document would only hold a reference to the file holding the Java class, which would be loaded and run by the Java Virtual Machine, incorporated in or linked to the browser software. The main advantage is that execution is safer and much faster compared to interpretation of JavaScript, because the code is already compiled. This set-up is shown in Figure 6.6.

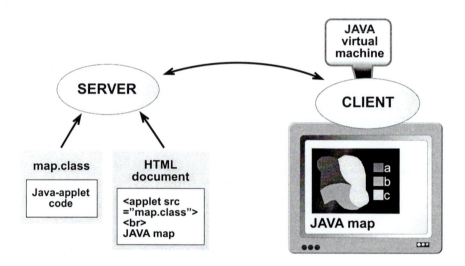

Figure 6.6 Set-up using Java applet.

The functionality offered by Java applets or scripts can vary from offering very simple button interactivity (using "roll-over" scripts to have a button change appearance when the mouse moves over it) to a sophisticated interactive mapping environment. An excellent example of the latter is the Descartes system, designed to help with exploratory analysis of spatially referenced data, for example statistical data of the municipalities of Overijssel province, as in *URL 6.7*.

Using Java to extend Web functionality has become very popular because it combines the advantages of both client-side and server-side solutions. The application is stored and maintained on the server, thus bypassing the need for the user to download updates. The platform-independent nature of Java ensures the application will run on virtually any computer and because the actual processing takes place at the client-side, the server load is kept low.

6.4.2 Server-side functionality

Besides the addition of functionality on the client side as described in Section 6.4.1, there is also the possibility for utility programs at the server side to be linked to the server software. This linkage is in most cases achieved using the Common Gateway Interface (CGI). This protocol defines the communication between the server software and the application, but also the way that the web browser can transfer information to the CGI application through the web server. The latter is achieved by adding commands and parameters to the URL, using a question mark to separate the CGI part from the rest and using ampersands to delimit parameters (eg. http://kartoweb.itc.nl?doThis&parameter1&parameter2).

CGI-compliant applications can be used for any number of tasks, for instance to provide database access or to customise information depending on user preferences. Thus, the user could get a map depicting the latest figures from a database, which can come from a remote server, visualised with the colours and classification the user has requested (see the example in Figure 5.3). The set-up of such a system can be seen in Figure 6.7.

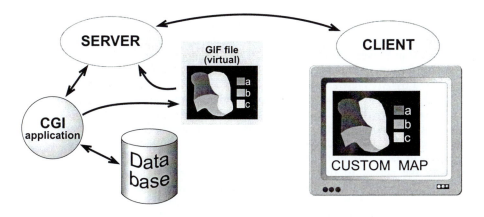

Figure 6.7 Server-side CGI set-up.

The nature of CGI-compliant software is varied: from simple self-programmed applications to support a page counter on a website"s home page to proprietary extensions to existing applications. Such an extension is the popular "Web Companion" plugin to Claris' FilemakerPro database software, which lets one publish an existing database on the Web. The much-acclaimed Oddens' Bookmarks site *(URL 6.8)*, that offers an extensive overview of links to cartographic websites, uses this software.

The cartographic functionality offered by such a system can vary from offering very simple static maps to a full-blown electronic atlas system, such as the

National Atlas of Canada Online (*URL 6.9*). This has evolved from using CGI software developed in-house, as in 1993 when it was started there was no "off-the-shelf" software available for interactive mapping on the Internet. The software for the most recent 6th edition was programmed using ESRI technology (Frappier & Williams, 1999).

The advantage of server-side solutions is that the system can be used by any Web browser on any operating system without the need to install plugins, thus maintaining platform-independence. The main problem is the server load. On popular sites, the number of requests (usually called "hits") can run up to hundreds every minute. This puts tremendous demands on the applications running on the server side. The web server software itself has the comparatively easy task of receiving requests and sending files in response, but imagine a database system having to keep up by processing a hundred SQL-queries per minute!

6.4.3 Mixed solutions

Although the previous paragraphs separated client-side from server-side solutions, in practice often a combination of the two is used, thus combining the advantages of both. An example of this is the set-up shown in Figure 6.8, as used by ESRI's ArcView Internet Map Server. ESRI has developed solutions to enable its GIS and mapping applications to publish on the new medium. Almost all major GIS vendors have done this and although their approaches differ in detail, most use the combination mentioned above.

As can be seen in Figure 6.8, the ArcView program takes on the role of a CGI application. For this purpose an ArcView extension is installed, called Internet Map Server (IMS). This enables CGI-compliant commands to be received from the browser, through the web server. The command, say a map query, is processed by ArcView and the result (a map view) is converted to a GIF or JPEG file and sent to the browser. A Java applet called MapCafé is used to implement in the browser an interface similar to the standard ArcView interface. Clicking the zoom button and dragging a rectangle in the map would result in the Java applet building a CGI command to implement the required zoom action. The last item in this set-up is a plugin (called esrimap.dll) to the web server software, which enables the server to find the appropriate ArcView application to handle the request. This application can run on another computer and to decrease server load the plugin can distribute requests among a multitude of computers running the same ArcView application. Because the standard ArcView application is used, existing projects can be made to run on the Web with little effort, as has been done in the example of *URL 6.10*, which shows the countries where ITC Cartography consultancy has taken place in the period 1982 1998. The Java applet can be customised and the IMS can handle calls to all functionality within ArcView, including its built-in scripting language (Avenue). This makes the system very flexible.

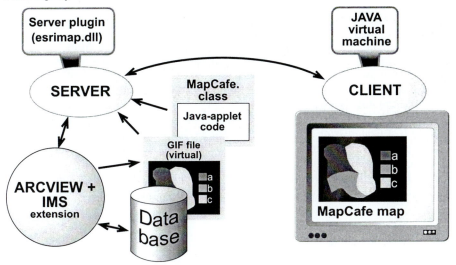

Figure 6.8 ESRI's ArcView Internet Map Server set-up.

In conclusion, it can be said that the Web is a powerful new way of publishing cartographic information. It offers a very simple mechanism to provide many users quickly with simple static maps at low costs. When the mapping environment needs to be more versatile, supplying interactivity or dynamic maps, functionality can be added on the client side or the server side. This could range from the simple addition of plugins to enable additional file formats, to complicated set-ups involving programming of additional Java or CGI applications.

URLs

URL 6.1   A Brief History of the Internet
            <http://www.isoc.org/internet-history/brief.html>
URL 6.2   PC Magazine's Beginner''s Guide to TCP-IP
            <http://www.zdnet.com/pcmag/issues/1520/pcmg0030.htm>
URL 6.3   A Short History of Internet Protocols at CERN
            <http://wwwinfo.cern.ch/pdp/ns/ben/TCPHIST.html>
URL 6.4   WWW Frequently Asked Questions <http://www.boutell.com/faq>
URL 6.5   W3C–The World Wide Web Consortium <http://www.w3.org/>
URL 6.6   Client-Side JavaScript Guide
            <http://developer.netscape.com/docs/manuals/js/client/jsguide/index.htm>
URL 6.7   Explorative maps of Overijssel using the Descartes system
            <http://allanon.gmd.de/and/java/iris/app/itc/indexm.html>
URL 6.8   Oddens' Bookmarks <http://oddens.geog.uu.nl>
URL 6.9   National Atlas of Canada Online <http://atlas.gc.ca/english/>
URL 6.10 ITC Cartography consultancy 1982-1998
            <http://kartoweb.itc.nl/avmaps/consultancy98/consult98.html>

REFERENCES

Frappier, J. and Williams, D., 1999, An overview of the National Atlas
  of Canada. In: *Proceedings of the 19th International Cartographic Conference
  ICC99, Ottawa*, edited by Keller, C.P. (Ottawa: Canadian Institute of
  Geomatics), pp. 261-267. (See also: <http://atlas.gc.ca/english/about_us/
  index.html #overview>)
Peterson, M.P., 1996, Cartographic animation on the Internet.. *Proceedings of the
  joint ICA Commissions Seminar on Teaching Animated Cartography*, Edited by
  Ormeling, F.J., Köbben, B and Perez Gomez, R (Utrecht: International
  Cartographic Association), pp. 11-14.
Salus, P.H., 1995, *Casting the Net: From ARPAnet to Internet and beyond*,
  (Reading, MA: Addison-Wesley Publishing Company).

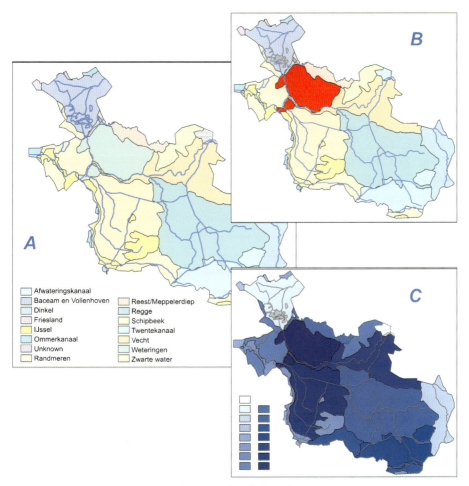

Afwateringskanaal
Baceam en Vollenhoven
Dinkel
Friesland
IJssel
Ommerkanaal
Unknown
Randmeren
Reest/Meppelerdiep
Regge
Schipbeek
Twentekanaal
Vecht
Weteringen
Zwarte water

Coloured version of Figure 5.8.

Coloured version of Figure 7.13.

Plate 1

Web Safe colour palette (1)

The colour codes are short versions of HTML (hexadecimal).
The codes can be related to intensity values of Red, Green
and Blue on a decimal scale of 0 to 255.

Hex.	Dec.	Example
00	0	
33	51	6F9 in disc = #66FF99 (HTML)
66	102	= 102R + 255G + 153B
99	153	
CC	204	(In the colour disc above, one or more
FF	255	of RGB has hexadecimal value FF)

In this and the following plate, **hue** is determined by
direction, **saturation** increases outward and **lightness**
either decreases or increases outward.

Plate 2

Web Safe colour palette (2)

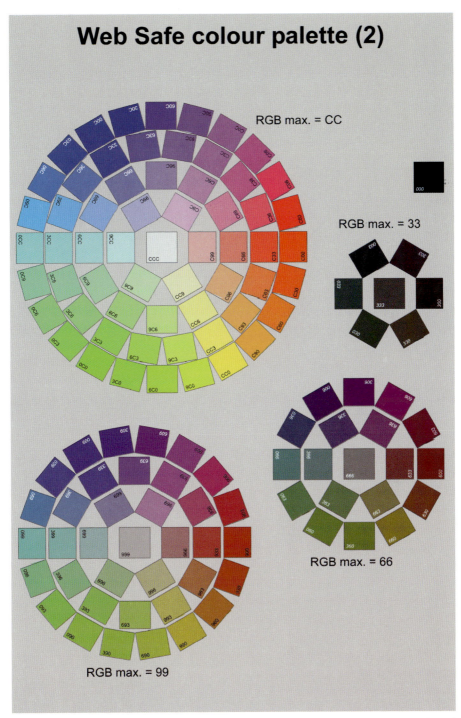

RGB max. = CC

RGB max. = 33

RGB max. = 66

RGB max. = 99

Plate 3

Coloured version of Figure 7.14.

Plate 4

Coloured version of Figure 7.17.

CHAPTER SEVEN

Web map design in practice

Jeroen van den Worm

7.1 INTRODUCTION

Designing maps for the Web is a new and challenging task. The cartographer must consider the limitations and opportunities offered by on-screen maps in general and by the special characteristics of the Web in particular. Even without using the Web, on-screen maps offer more facilities than paper maps, of which interaction is the most prominent. Furthermore, on-screen map design is no longer restricted by factors related to physical map reproduction, although the cartographer needs to bear in mind that some users may want to print web maps. Of course disadvantages also exist as has been explained in Section 4.4.

However, when designing maps for use on the Web, even experienced cartographers have to adjust their map design habits towards the nature of the Web and what it can offer. Not every map design that is successful on, for example, a CD-ROM will be equally successful when sent over the Web. Web map designers have to worry about factors over which they have little control, and which are influenced by user actions and system configurations. These include fonts and colours. Designers also have to keep file size small to speed up downloading. This could lead to smaller maps and simpler designs than the cartographer would prefer. However, the possibility to add links or apply mouse-over (or roll-over) techniques to these simple maps can overcome most disadvantages.

This chapter deals with the practical application of basic cartographic principles as discussed in Chapter 5. New technology opens the way to the application of other "derived" graphic variables such as shadow and transparency, which were infrequently or not at all applied in conventional map design. Special attention will be paid to web design options such as the application of so called web objects. These include all interactive "gadgets" such as pull-down menus, mouse-over events, hotspots, etc. that invoke action.

7.2 FACTORS INFLUENCING MAP DESIGN

Chapter 2 dealt with the cartographic visualisation process, guided by the phrase: *"How do I say what to whom and is it effective?"* This should also be applied to maps on the Web. However, the last part of the phrase will not always be unambiguous, especially when new technological options are used, such as mouse-over or animations, the impact of which is not yet fully understood.

Revisiting Figure 1.2, web maps are divided into static and dynamic maps, each subdivided into view only and interactive.

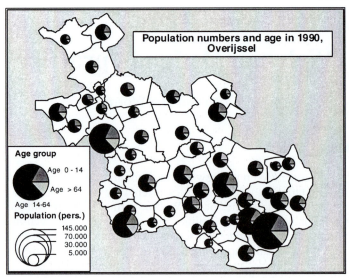

Figure 7.1 A view only map: the population age characteristics of Overijssel.

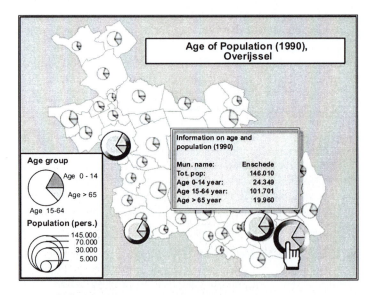

Figure 7.2 An interactive map with access to more detailed data.

Maps of the view only type will fit the "presentation" approach as shown in Figure 2.2. The cartographer processes the data ("*what*") and presents it via a specific cartographic method and map symbolisation ("*how*") bearing in mind a specific user group and expected map use ("*whom*"). "*Effective*" can be judged based on experience and previous map user research. Considering the role of the web cartographer the design approach to these maps is not much different from traditional map design, if one indeed incorporates the typical web characteristics treated in the next section of this chapter.

The cartographer still has a considerable role to play in the design of those interactive maps with which the user is able to interact in limited, pre-defined ways, e.g. by making some symbols "clickable" to reveal more information. These maps still fit the "presentation" approach. The basic differences between the view only and interactive map types lead to a different design approach for each type.

1. Map symbols in static or dynamic view only maps can be compared with map symbols in conventional maps. Their design should be based upon the nature and required perception of the data to be visualised. An example is found in Figure 7.1. It shows population data for the municipalities of the province of Overijssel in the Netherlands. Size has been applied to depict the quantitative character of the information to be displayed, while colour has been used to express certain qualities.

2. Map symbols in those interactive maps where "mouse-over" or "clicking" leads to further information or data, are not per definition related to the nature of the mapped data (*URL 7.1*). In both cases it is the navigational, interactive function that defines their design. In Figure 7.2 the map from Figure 7.1 has been redesigned and some symbols now express the idea: "Here you can find the information you are looking for." When these symbols are activated by "mouse-over" a small table appears with more details on the mapped subject. Figure 7.3 shows a map as well but the aim of this image is not so much to depict data directly but to be attractive and to invoke action to access other information (*URL 7.2*).

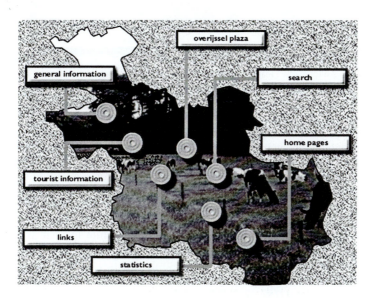

Figure 7.3 An interactive map as index to other information.

If interactivity is taken a stage further, the role of the cartographer may be less sharply defined. He may simply be the person who provides cartographic symbolisation "tools", so making it possible for users to make maps based on their own data and to use the maps to explore datasets. Here it is the user who ultimately decides which design options will be selected and how these will be applied. Referring to the above phrase *"How do I say...."* it can be said that *"I"* and *"whom"* have become the same person. If the user is only interested in data exploration (see Figure 2.2) he could of course deliberately ignore established cartographic symbolisation and perception rules in order to emphasise certain aspects of the data. Figure 7.4 gives an example.

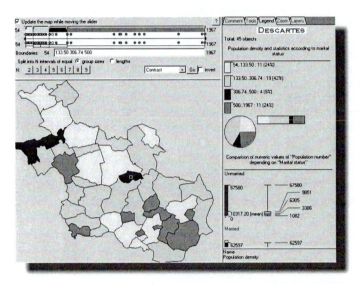

Figure 7.4. An interactive map for exploratory purposes (using Descartes software, *URL 7.3*).

The examples illustrate that the function of the web map has a direct impact on the map symbol design. Besides this, it is important to keep in mind that not all visitors to a page containing a web map are per definition interested in the map and/or its contents as such, as was explained in Chapter 3. Perhaps they did not specifically look for a map, but the map was found on their way, "surfing" the Web and "zapping" from page to page. To hold the user''s attention such a map should not only be functional but also be attractive from an aesthetic point of view. In addition, it is important for the user to discover very quickly whether a map is interactive or not. Therefore, successful web maps have to be specially designed for their purpose (*URL 7.4*). It is for example worthwhile to visit and critically study the maps found on the sites mentioned in Chapter 9, on web maps and tourists. Some of these maps clearly demonstrate that just using a scanned version of an existing paper map (even if it is well designed) does not necessarily result in a successful web map (*URLs 7.5, 7.6 & 7.7*). Combining functionality with a high level of visual attraction and a design that suits the medium should be the challenge for every web cartographer.

Factors involved in map design include the analysis of geospatial data characteristics, the definition of map content and the required perception levels to portray the data based on the purpose of the map, the existence of map symbol associations and standards, the scale and required accuracy. Most of these factors interrelate and influence each other. As in all mapping, aesthetics, production costs and time play a prominent role as well. When designing for maps to be displayed on a screen (including web maps) special design opportunities arise because of the additional properties of the medium compared to paper. Interactivity, animation and multimedia (sound and video) spring immediately to mind here. Maps on screen can also make use of graphic and symbol design options that, in the days of analogue paper map production, were not possible, or possible only with great effort. However, as has been mentioned, various limiting technical factors come into play with respect to web maps because of the specific character of the Web. One of the biggest limitations is the frequent need to keep map size small in order to facilitate downloading. This need in turn leads to the need to limit the amount of information on a map and to take great care with generalisation. However, it is worth noting that technical developments might solve some of these problems. For example, much information can be carried even in a small file by the use of the vector approach to web map design (see Appendix A).

Because of the situation sketched above, the visual hierarchy of web map information content deserves more attention compared to maps in general. Three distinct levels can be distinguished (Figure 7.5).

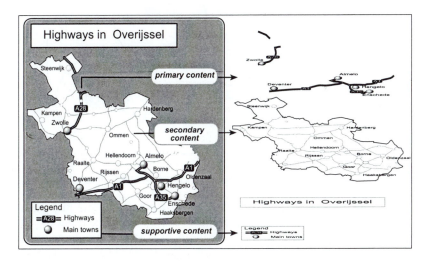

Figure 7.5 Web map content levels.

- The primary content level is formed by the main theme of the map. Interactive web objects such as hotspots, mouse-overs, etc. that will trigger specific events resulting in the supply of main theme information can also be considered to form part of the primary level.

- The secondary content level refers to the (often topographic) base map information, but also to pop-up menus, movies, tables, sounds, etc. supplying additional information on the main map theme.
- The supportive content level includes marginal information such as legend, grid, illustrations, graphs or for instance pop-up menus supplying the user with information that is not directly related to the main theme of the map.

The content of the map depends also on the scale of the map. In principle maps on screen, and thus web maps, have no fixed scale, since they can be enlarged or reduced at will by "zooming" in or out. This changes the relation between the displayed map distances and real world distances, and thus the scale. There is, however, an ideal scale (or scale range) to display any particular map, depending on the density and accuracy of map detail. If a map is enlarged too much very few details may be visible in the image window and the positional accuracy of the symbols may be much less than the user expects from the scale. If the map is reduced too much it will become illegible. The option to zoom with web maps depends on the system and the plugins installed, and upon enough detail being supplied to allow considerable enlargement. Web cartographers can use three distinct zooming strategies or options.

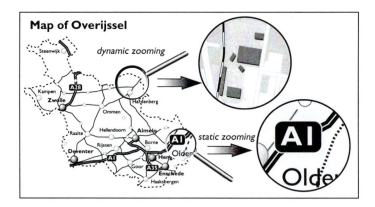

Figure 7.6 Static and dynamic zooming of a vector-based image.

Static linear zooming. The relation between zoom factor and map content is static. When zooming into the map, the image is linearly enlarged but the content of the map does not change. In this case the map is stored simply as an image. Vector-based images (such as those stored in SWF or SVG format, see Appendix A) will keep their sharp character when enlarged but raster-based images (such as those stored in GIF or JPEG format) will show enlarged pixels (see Figures 7.6 and 9.6). This is the most commonly applied zooming option.

Static stepped zooming. As for the previous option, but in this case a series of maps of the same area is available, each one designed for a different scale or scale range. When the user requests to zoom in or out, the software automatically chooses the most suitable map for the desired scale. This system is widely used on route planning sites and by companies such as MapQuest (*URL 7.8*). A variation on this technique is to allow a "magnifying glass" to be moved over a map, with a large-scale version of the map appearing under or near to the glass (*URL 7.9*).

Dynamic zooming (animated scaling). In this system there is a direct relation between scale and map content. The larger the scale the more detail is shown in the image. A direct link between the image and some kind of database is necessary. Although not always required, the cartographic symbolisation may change with scale. For instance a town represented by a point symbol at a small scale may turn into an area symbol upon zooming into the map (*URL 7.10*).

If scale is considered to be important for a web map it is necessary to include a scale bar because the actual size of the map, and therefore its scale depends on the user's system configuration. This will enable the user to measure the length of the line on the screen and to calculate the scale of the map. The cartographer must decide on a "default scale" for the map at a specific default display size. This scale can then be used as a reference to define map content and the required accuracy. On-screen maps, just as their paper counterparts, should be characterised by the positional accuracy required by the nature of the spatial data to be displayed. For a web map primarily intended as an interface tool for linking to other information, absolute positional accuracy is of less importance although relative position should be maintained. In addition to positional accuracy, semantic accuracy is just as important in web maps as in other maps.

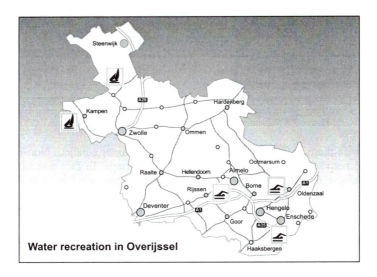

Figure 7.7 Web map without a legend.

Semantic accuracy is related to the way that data are symbolised and represented. These in turn are related to the intended web map use. The semantic accuracy of the map symbols depends upon the appropriate selection of the graphic variables and their variations. Web maps may be more limited in this respect than paper maps. One reason for this is the assumption that in general the viewing time for web maps is short in comparison to their paper counterparts. In addition, the information content level of web maps is lower because of low

7.3 CARTOGRAPHIC SYMBOL DESIGN FOR WEB MAPS

Over the centuries cartographers have developed a very wide variety of symbols to portray different kinds of information. On web maps, however, these symbols may be used differently than on paper maps.

7.3.1 Point symbols

Point symbols have several different uses on maps. Examples include: depiction of geographic features that occupy a very small area on the map; representation of data referring to a geographic unit (for example the proportional circle symbols of Figure 7.1); providing "shorthand" information, e.g. an array of symbols next to a town on a tourist map to show the tourist facilities available. On maps one distinguishes three categories of point symbols: pictorial, geometric and alphanumeric (see Figure 7.11). On web maps a point symbol is often also a web object, that is an area that can receive mouse "events", enabling JavaScript functions and hyperlinks (*Using Fireworks*, 1999).

Due to the typically low resolution of web maps, it may be difficult to design complex point symbols. Because of the special capabilities of web maps this is not necessarily a drawback since web objects can be used to access the second level of information content. Since web maps may attract inexperienced map readers, web cartographers often attempt to use pictorial point symbols (see examples in Figures 7.7 and 7.11), despite the practical difficulties involved. In general such symbols are easy to understand, perhaps even without the use of a legend.

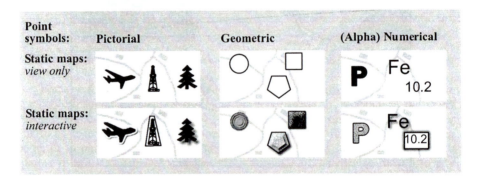

Figure 7.11 Point symbols.

The main problem in designing pictorial symbols for web maps is that within a small area, actually a limited number of screen pixels, the essential characteristics of the phenomenon must be visualised. These symbols are best used to represent qualitative data. They should be uncomplicated and easy to understand. If different symbols are applied on a map, they all should have the same visual impact. For on-screen maps these symbols may need to be larger than on the equivalent paper maps in order to aid legibility. For all these reasons the creation of pictorial symbols can be time consuming. For those who are less creative or who have less time available, software packages for web design, general graphic design and GIS offer clip-art libraries containing such symbols free of copyright (*URLs 7.16, 7.17 & 7.18*). Most of the symbols offered in these libraries are however not suitable for cartography, often being too large and complex. Yet, with some common sense and a feeling for visual balance, it is possible to collect a library of useful symbols. It is advisable to keep in mind the relation that should exist between the symbol and the user's real world knowledge. For instance, a simplified representation of a rice plant may mean nothing to a northern European.

Geometric or abstract symbols do not attempt to resemble the real feature represented. On different maps the same symbol can have a different meaning. Therefore geometric symbols should always be explained in a legend. As has been mentioned earlier, this may create specific problems for web maps, where finding space for a legend may be difficult. Web design software often offers basic drawing tools that enable fast and easy construction of geometric shapes. However, it is also possible to install and use system symbol fonts such as Zapf Dingbats and Wingdings. The advantage of geometric symbols over pictorial symbols is that their size can be relatively small. These symbols can be easily varied in size to represent quantities, while their shape and colour can convey qualitative information.

The third group of point symbols uses letters and numbers. Although this type of symbol has some disadvantages such as the fact that distribution patterns may not be immediately obvious and a legend is required to understand their meaning, they are found on many maps. Especially for web maps they have to be made relatively large to be effective (*URLs 7.19 & 7.20*).

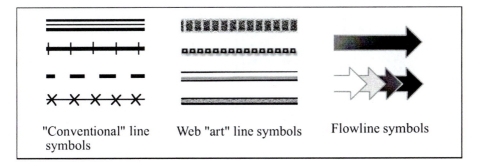

"Conventional" line symbols   Web "art" line symbols   Flowline symbols

Figure 7.12 Line symbols.

7.3.2 Line symbols

Line symbols on topographic maps are used to represent features such as roads, railways and contours. They are also widely applied in thematic cartography, for example to show the position of geological fault lines, ocean currents, trade flows or average temperature (see Figure 7.12). They can also be used in more abstract ways, for example to visualise geographic links between URLs. When designing thin and/or highly curved line symbols, the cartographer should be aware that some specific graphic variables such as orientation and/or texture are less suitable, especially considering the limited capabilities of the Web. In web maps, line symbols may be animated, for example to present traffic flows. As for point symbols, clip-art line symbols can be used. The thin and elongated shape of line symbols makes them awkward to handle as interactive web objects, especially when they are highly curved, because web objects have to be defined as areas (polygons).

7.3.3 Area symbols

Area symbols are applied to present area-based information. The graphic variables typically used in designing area symbols for maps are colour, value, texture, shape and orientation (Figure 7.13, coloured versions in Plate 1). Many web design programs offer tools to apply these variables to areas. Taken together, these variables can be used to create quite complex area patterns. Combining the variables can have several aims: to promote the semantic meaning of the symbol (for example green tree symbols to represent forest areas), to decrease the possibility of confusion among symbols, to accentuate the figure-ground relationship or to contribute to the aesthetic impression of the map. The careful use of variables other than just "flat" colour and value can make a web map more lively and interesting, though it will often have the effect of making the file size larger and downloading slower. In web maps the area symbols can function very well as web objects, because in general they are relatively large (*URLs 7.21 & 7.22*).

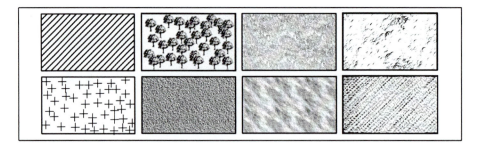

Figure 7.13 Graphic variables applied to area symbols.

7.3.4 Combination of symbols in a map

Map design is more than the design of the individual symbols. It involves all information to be represented in the map simultaneously. As described in Chapter 5, this is in principle a rather straightforward intellectual process consisting of a series of steps that should follow each other logically, beginning with an analysis of the type of information to be presented. Anybody following these steps should be able to produce a functional map, of course also paying attention to the overall layout. Someone with artistic talent may be able to make the map more interesting and exciting to look at while still keeping to the cartographic principles, but this extra artistic contribution is not really essential.

Figure 7.14 gives an example of a map designed according to the principles described in Chapter 5. It is a tourist map of Overijssel for the provincial web site and is intended to give the user an overview of those places that are interesting to visit. It is obvious that the map should also have a promotional function, so attractiveness has a high priority on the wish list. The map itself should therefore make the province look attractive to visit (for the full effect, see the coloured version on this book's website, and in Plate 4).

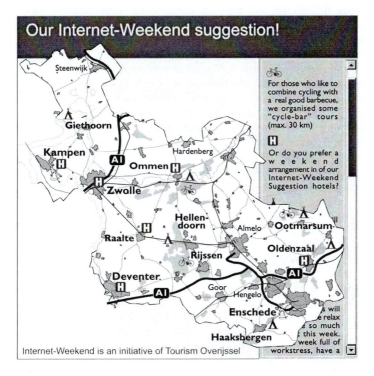

Figure 7.14 Tourist map of Overijssel designed according to the principles of Bertin.

The content of the map shows detailed information on the road network, settlements, etc. A point to note is that the map has been designed as if it were view only, with short descriptions of what each symbol signifies. Some of the point symbols are in fact clickable but the design does not focus on this functionality. Therefore, in order to find this out the user must explore the map by randomly "scanning" the map for clickable objects. It is possible to improve the design to draw special attention to these, as will be discussed in Section 7.6 below.

7.4 COLOUR ON WEB MAPS

The technical aspects of sending and receiving coloured images over the Internet are discussed in detail on many websites (e.g. *URL 7.23*). The important thing that has to be borne in mind is that the sender has no direct control over how the image will appear on the user's monitor: it all depends on the particular configuration and settings. For this reason, web map designers often adopt a cautious approach and assume the minimum configuration and lowest settings. Although most computers nowadays are able to display 16-bit or 24-bit colour, displaying over 65 thousand or over 16 million colours respectively, it is safest to assume that the user's configuration is set to 256 colours. The 216 colours of the Web (or Browser) Safe colour palette fit into this and are guaranteed to be "non-dithered" (see Appendix B.2) on any configuration (see Plate 2 & 3). This does not mean, however, that these colours will always appear exactly the same on any system, since much depends on the calibration of the monitor.

For a map using only "flat" colours, like the map in Figure 7.14, it is unlikely that more than 216 different colours will be needed, so the use of the Web Safe palette would appear to pose no problem. This is not entirely true, however. Map designers often like to make use of very pale colours, especially for large areas that need to be treated as background, yet the Web Safe palette offers very few pale colours. In RGB terms, pale colours are achieved when all of these have values fairly close to 255 (on an intensity scale of 0 to 255). The Web Safe palette, however, offers nothing between 204 and 255. Many map designers just ignore the problem and simply choose pale colours from a larger palette, in the hope that the users also will be able to use a large palette.

Maps on screen can be made to appear more interesting and attractive if colours are used in a more artistic way, with for example colours blending into each other or fading round the edges, or by the use of transparency and shadow effects (see the following section). The graphic software packages available today support many such options. Furthermore, the cartographer may wish to include photographic images as an integral part of the map. Making use of these possibilities will in most cases lead to images containing more than 256 colours, automatically leading to dithering in configurations set to 256 colours (see Appendix B.2). What also needs to be borne in mind is the file format used to compress the image for sending over the Web.

The common file formats GIF and JPEG used to compress raster images are described in Appendix A. GIF is suitable for an image containing up to 216 flat colours, so the GIF compression is ideal for maps such as the map in Figure 7.14. The colours are specified exactly, so the designer has good control over them, with the proviso that non-Web Safe colours will appear dithered on 256-colour configurations. JPEG does not work with a look up table, it compresses images based on colour and intensity. Even if one designs a map containing flat Web Safe colours, some of them may "shift" slightly during compression, leading to a dithered result on 256-colour configurations. For maps making considerable use of colour blends, such as the map in Figure 7.16, JPEG is however the best compression method. Web map designers are well advised to test both methods on each map they produce, varying the various parameters available. Most web design software packages offer tools to preview the results of different resolutions, colour palette sizes (GIF) or compression quality (JPEG). The aim is to find which gives good results (usually assuming a 256-colour configuration) while maintaining a reasonably small file size. The chosen result can be sent over the Internet as a final check to a minimum configuration in one's own office.

7.5 TYPE FONTS AND NAME PLACEMENT ON WEB MAPS

Robinson *et al.* (1995) state that some cartographers have claimed that names on maps are a "necessary evil". Nothing can be truer than this statement applied to web maps. Text on web maps cannot be omitted, as text can express information such as geographic names, height values, etc. that is not possible using any other graphic symbol. The impact of text on the appearance and perception of maps is considerable. Inappropriate selection and application of typographic variables such as font size or font variable may clash with the graphic variables such as applied during the cartographic symbol design process. Therefore, map typography and map symbol design cannot be separated from each other. In general, one can state that the perception characteristics of the graphic variables as described in Chapter 5 can also be applied to type. Size refers to type size, expressed in points. Shape refers to variations in font (or type face) for instance: serif types (Times, Bodoni, Garamond, Gill) or sans serif types (Helvetica, Univers, Tahoma). Some cartographers consider also the difference within one font such as lower and upper case as a shape variation. Orientation refers to upright or italic variations within one font. Value refers to the font variations: light, medium and bold, but may also refer to a graded grey value range applied to a font or fonts. For the practical application in maps of the typographic variables, see Figure 7.15.

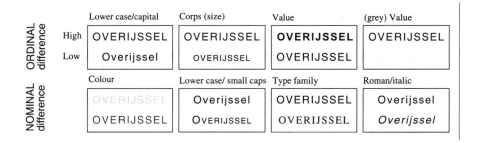

Figure 7.15 Typographic variables and their map application.

Considering text on maps on the Web one can distinguish two main applications: text applied outside the map face, such as in the legend, scale line, title, grid, etc. and text within the map face. Text outside the map face can be treated like any other text application within a web page. This includes precautions taken to avoid font replacement or to control text flow, letter spacing and leading. Within the map face, text is in general applied for geographical names, as symbols or for a limited amount of descriptive information. Here also web-related constraints should be taken into consideration. To be sure that no text changes will take place at the user's side the map is best saved in raster format, including the text. If text is sent separately (defined as vectors), it is best to use a very common font that the user will have installed as a standard font or to include in the HTML code an alternative font. If a less common font is used there are two options: either to generate a GIF image of each word and place it on the web page or to "embed" the font in the web page, i.e. to send the font description file to the user (this may require a licence).

The readability of text within the map face of a web map is influenced by several major factors (Ditz, 1997): selected font, font variation and font size, font orientation (upright, italic), text placement (horizontal, inclined or curved along a path), figure-ground relation and the amount of anti-aliasing (see Appendix B.2). Some fonts produce more readable text than others on web maps, especially at small sizes. Simple styles and so-called "open" fonts are particularly suitable. The selection of the bold text version may improve readability but this may sometimes clash with desirable perception characteristics, when for example the cartographer would prefer to use light text for less important features. On web maps capital lettering may also improve readability, particularly at small sizes. This is in contradiction to paper maps, on which lower case lettering is generally preferred.

Text size is traditionally expressed in points. However, 10-point text in one particular font may appear somewhat larger or smaller than 10-point text in another font. Furthermore, text size on a screen depends on the monitor resolution settings and on the computer platform and operating system. When text is rasterised, as it often is for transfer over the Web, the size of the pixels in relation to the text is very important. The general rule is that text to be rasterised should not be smaller than 10 point.

Upright and italic text are often used together on maps, for instance to depict differences between man-made and natural features respectively. The sloping lines of italic text may appear very "jagged" on typical low-resolution web maps. Because italic text is often applied to name natural features, it is also very often curved along a path, following the extension and direction of the feature. This combination may make the result even worse. Even upright text used for inclined names may become less legible. It is obvious that this constraint has a big impact on the placement of, for instance, river names or the names of streets in a street plan. There are a number of possible solutions to minimise these negative effects, for example the selection of open letter fonts, the application of a moderate anti-aliasing rate and the avoidance of extreme inclines. Note in this regard that a user of a web map cannot rotate it to read very inclined, even upside-down text, as the user of a paper map can. For curved names, such as rivers, upward curves are more legible than downward curves because the lower case Roman alphabet has more ascenders than descenders. Readability can also be improved by giving 10 to 20% extra character kerning.

The figure-ground relation and anti-aliasing rate also affect the legibility of the text. The figure-ground relation between text and background can be improved by adding (if possible) a thin white outline to the text but the selection of light type against a saturated or "noisy" background is best avoided. Another option is to apply (transparent) cast shadow. This option is best applied sparingly and just for important text, as cast shadow also creates "noise" and may increase file size. The application of anti-aliasing also tends to increase file size, but a moderate antialiasing rate improves readability, particularly for curved or inclined text. This applies also to point symbols selected from a type font such as Wingdings.

What is said here about the application of type in maps does not guarantee a good result: the finished map should be tested on the expected minimum user configuration. Even type selections which have appeared to work quite well on some maps may not give good results on others. A good deal depends on the actual density of text. If very much text must be placed in the map, the application of a control-panel legend enables the user to switch text layers on or off. This allows the designer to use larger, more readable type sizes.

7.6 ADVANCED DESIGN

The graphic variables described in Chapter 5 have become widely accepted since they were introduced at the end of the nineteen sixties by Bertin. Bertin was mainly concerned with monochrome maps, so he treated colour as a single variable. In fact, however, colour can be considered to have three variables: hue, chroma (or saturation) and value (or lightness). The last of these is of course already included in Bertin's original list. All of these three variables of colour are incorporated in colour models such as Munsell or, for use in the computer, HLS. Hue is then equivalent to Bertin's use of the term "colour", while the additional variable saturation can be used, for example, to express an ordinal quantity. In addition to splitting colour into its three variables, over the years since Bertin published his original work several authors have suggested additional variables or variations on the existing variables (MacEachren, 1994). For example, several variations of texture have been proposed based on directionality, size and density of the patterns. Applying the newer variations in combination with the traditional graphical variables leads to some interesting solutions. Two effects will be elaborated upon here: transparency and shading/shadow.

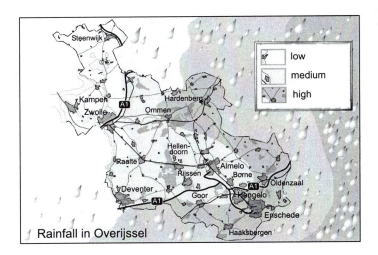

Figure 7.16 Application of transparency: a rainfall map of Overijssel.

Transparency represents a kind of fogginess, by which for instance map themes are obscured or faded in proportion to their certainty. In practice, this is achieved by reducing the value and saturation ranges (i.e. reducing contrast) and allowing all colours to approach a light grey. The term transparency is also used to describe map features that overlap each other whereby the feature "underneath" is dimly visible under the feature "on top". Transparency in the sense of fogginess is an especially useful variable. For example it can be applied to subdue the background in a map in order to enhance the main theme in the foreground. This variable, like the others, can of course also be applied in other visualisation techniques apart from standard maps. For example, the impression of depth or distance in an oblique view can be enhanced by increasing the fogginess towards the back. Of course this variable must be used with care – too much fog and nothing can be distinguished. Figure 7.16 shows the application of the variable transparency in a web map.

Shading and cast (or drop) shadow can be used to enhance the contrast between "figure" and "ground". Several web design programs offer shading and cast-shadow (or drop-shadow) options. Both of these increase the sense of depth. The shading of map objects can be compared to the creation of hill shading. The colour of the horizontal surface is taken as the starting point. Then all sloping surfaces are made lighter or darker according to the vertical illumination angle of the imaginary light source, its direction and intensity. Conventionally the illumination direction is from the NW, as applied for hill shading. To create the impression of a depression rather than a hill, one does not actually need to make a new 3D model. It is sufficient to move the illumination direction to the SE. Shading is widely applied in web map design to create "3D" clickable objects that appear raised before clicking and depressed when clicked (see Figure 7.2).

Cast (or drop) shadow is caused by the object itself casting a shadow on to the background. It is possible to use this to create the impression of objects (symbols) "floating" above the background. The apparent vertical separation between object and background depends on the "offset" and elongation of the cast shadow, and also on the sharpness of the shadow edges. The impression of greatest distance occurs when the offset is large, the elongation small and the shadow edges rather blurred. The visual effect of cast shadows is not so dependent on the direction of illumination as is the case for shading, but NW illumination is mostly used (*URL 7.24*).

The design of the tourist web map of Overijssel shown in Figure 7.14 is suitable for a view only map. Furthermore, the symbol design is restricted to the original graphical variables of Bertin. The full potential of putting maps on the Web is, however, only achieved when they are made interactive in some way. This added functionality can be conveyed to the user by using some of the additional visual variables just mentioned, to draw attention to clickable objects and areas where "mouse-over" applies. In this context buttons or symbols are often made to appear raised or depressed by the application of shading. In order to convey the impression of a very general, inexact location of symbols, especially point symbols, cast shadow can be used to make them appear to "float" above the background.

Tourist web maps are typically used to guide the general public looking for interesting places to spend some free time, so ease of use and attractiveness are highly desirable qualities. The map in Figure 7.14 could be criticised on two counts: clickable objects and hotspots are not immediately obvious, and the design is rather conventional and not very exciting. The first criticism can be met by improving the figure-ground relation and by emphasising clickable objects and hotspots by the application of the design tricks just mentioned. This will make the map as a whole more interesting (Figure 7.17 and Plate 4). Forest is shown now by a combination of colour (only on the web version), form and shading, resulting in a natural-looking texture. Shading has been applied to create a semi-3D effect to the water bodies and forest areas.

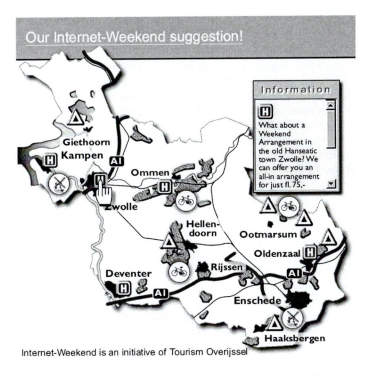

Figure 7.17 Tourist map of Overijssel specially designed for the Web.

The figure-ground relation is emphasised by applying a drop-shadow effect to the provincial boundary. Shading is also applied to create a raised effect to the point symbols, emphasising their clickable character. By selecting one of these objects the symbol appears to become depressed and a window pops up giving information about (in this case) the current tourist weekend arrangements. These symbols are connected to a database that is updated weekly. The cyclist symbol can be clicked to link the user with a downloadable and printable detailed cycling route map. To reduce download time and promote a shorter map reading time the content of the map is simplified to a bare minimum.

Many of the remarks on design found in this chapter can be applied to animations as well, especially since most animations used on the Web are composed of a set of individual (bit) maps. Each individual frame of an animation is best kept rather simple in order to allow the viewer to concentrate on the main theme. In fact, not overloading maps with too much content is a basic principle applicable to all web maps.

URLs

URL 7.1   Mouse over <http://www.waterland.net/rikz/waterstand/index.html>
URL 7.2   Mouse over <http://www.waterland.net/>
URL 7.3   Descartes, interactive maps
            <http://borneo.gmd.de/and/java/iris/app/itc/indexm.html>
URL 7.4   Map design <http://www.intermaps.com>
URL 7.5   Scanned map on the web: Stanford University, USA
            <http://www.stanford.edu/gifs/campus.1620-63-4.gif>
URL 7.6   Scanned map on the web: SFO – student housing, the Netherlands
            <http://www.student.wau.nl/asserpark/plattegrond/index.html>
URL 7.7   Van den Broeke Real Estate, The Netherlands
            <http://www.lodging.nl>
URL 7.8   Static stepped zooming: MapQuest online maps
            <http://www.Mapquest.com>
URL 7.9   The Freehand Source, tips archive: flash magnifying lens effect
            <http://www.freehandsource.com>
URL 7.10 The geography network > view live maps > ESRI: examples of
            Animated scaling <http://www.geographynetwork.com>
URL 7.11 Legends: ABC News: World section
            <http://www.abcnews.go.com/sections/world/balkans_content>
URL 7.12 The geography network > view live maps > ESRI: examples of
            'hidden' legends <http://www.geographynetwork.com>
URL 7.13 Legends: University of Manitoba, Canada
            <http://www.umanitoba.ca/about/map>
URL 7.14 Layer legends <http://www.waddenzee.nl/english/frames.htm>
URL 7.15 Layer legends <http://www.census.gov>
URL 7.16 Clip art: Online graphic encyclopedia <http://www.symbols.com/>
URL 7.17 Free WebArt Clip library <http://www.iconbazaar.com/>
URL 7.18 Clipart Xoom Web Services <http://www.xoom.com/>
URL 7.19 Alpha numerical point symbols
            <http://www.utexas.edu/maps/main/overview>
URL 7.20 Alpha numerical point symbols
            <http://www.wordplay.com/stjohnsmap>
URL 7.21 Interactive area symbols German WWW Server <http://www.entry.de>
URL 7.22 Interactive area symbols <http://www.trabel.com/brussels-map-heysel.htm>
URL 7.23 The Browser Safe Palette <http://the-light.com/netcol.html>
URL 7.24 Tools for web design <http://www.sausage.com>

REFERENCES

Ditz, R, 1997, An interactive cartographic information system of Austria –
conceptual design and requirements for visualization on screen. In *Proceedings,
Volume 1, of the 18ᵗʰ International Cartographic Conference ICC97*, edited by
Ottoson, L., (Gävle: Swedish Cartographic Society), pp. 571-578.

MacEachren, A.M., 1994, *How maps work: representation, visualization,
and design*, (New York: The Guilford Press), pp. 270-276 and 370-376.

Robinson, A.H., Morrison, J.L., Muehrcke, P.C., Kimerling, A.J. and Guptill,
S.C, 1995, *Elements of Cartography*, 6ᵗʰ ed., (New York: John Wiley & Sons),
p. 404.

Using Fireworks 3, 1999, (San Francisco: Macromedia Inc.).

CHAPTER EIGHT

Web maps and
National Mapping Organisations

Rob M. Hootsmans

8.1 THE CHANGING ENVIRONMENT OF NATIONAL MAPPING ORGANISATIONS

Geographic core data form the framework for most geodisciplines. These data provide a base on which application-oriented data can be overlaid, or a frame to which they can be attached (*URL 8.1*). By tradition, National Mapping Organisations (NMOs) are responsible for a country's geographic core data. However, the leading role of NMOs is becoming less obvious as a result of unavoidable changes in their environment. One aspect of these changes is the growing interest in connecting and sharing distributed databases by means of a national geospatial data infrastructure (see Section 3.5). Data users with access to this infrastructure – presumably via the World Wide Web – will be able to compare the products of various data producers, for instance by cost or quality. This implies that the NMO core data are no longer the only authority in this area. Still, NMOs have an opportunity to anticipate these changes and to play an initiating role in setting up a clearinghouse for a country's base data. In this chapter it is investigated how well prepared the NMOs are for such a new role by evaluating their use of the World Wide Web, and the challenges that the WWW can offer. For a current overview of NMO websites, see *URL 8.2*.

8.1.1 Current role confronted with changes

In the early days of surveying and mapping the agenda was strongly determined by defence and public safety initiatives, demanding a full coverage of the country at specified map scales. Later this was followed by settlement issues, resource development and public infrastructure. Currently, the agenda is rapidly moving towards the development of customised applications and the response to other societal issues, such as environmental monitoring (McLaughlin and Coleman, 1999). The latter issue clearly involves processes beyond a country's boundaries, implying that NMOs should prepare for international collaboration.

Additionally, Rhind (1997) states that most National Mapping Organisations are in the midst of dramatic changes. These changes are guided by new technology, by government decisions (financial cut-backs, less direct influence, more finances from the private sector), and competitive market demands. The future of NMOs will mostly be determined by their response to these inevitable changes. The time period and final results of these changes are still vague, but the organisations have a future as long as they accept that they are living with uncertainty and will remain in a permanent state of change (Rhind, 1997), and act accordingly.

8.1.2 Call for customised NMO products

As one result of the process of change Morrison (1997) observed the considerable "democratisation" of cartography, which especially NMOs can no longer ignore. It is no longer only the map producer who dictates what is put on a map. More and more the map users are being equipped with the electronic tools to create visualisations of their geographic thinking, as has been explained in earlier chapters. This trend was observed in a slightly different context by Peuquet and Bacastow (1991) in their discussion of topographic mapping for the US Defense Mapping Agency. In addition to one supply driven and centrally controlled set of highly accurate, hard-copy products intended for multiple uses, there is a growing demand for separate, single-use sets of products at differentiated accuracy levels – determined by the appropriateness of the available electronic tools and purpose of use. These products can be delivered as hard or soft copy, derived from one or more electronic databases, serving specifically stated purposes. These deliverables can be products such as digital raster graphics, e.g. a scan of the paper map product, or as digital line or vector graphics, which may be used directly in a GIS.

8.1.3 Position in the Geospatial Data Infrastructure

Any organisation that uses geoinformation needs easy access to their own data as well as externally retrievable data by means of a solid GDI, following standard documentation guidelines to describe the essential meta-information. Obviously, such an infrastructure should provide easy access to geographic base data. The World Wide Web has become the backbone for most Geospatial Data Infrastructures, since it already provides a network among organisations and allows transport of large quantities of data.

A national GDI should basically provide access to a country's framework data which are necessary for any specific geographic application (Figure 8.1). Framework data consist of data on specific themes, such as soil, vegetation and property information, as well as the core data, without which no sensible georeferenced analysis would be feasible. Core data include, for instance, the geodetic control network, ortho-rectified imagery, an elevation model, basic topographic information, information on administrative units and geographic names.

One can ask the question in which way the NMOs – as the traditional providers of the geographic core data – currently offer the basic data that users of spatial data need or require for processing in Geographic Information Systems. And additionally, do NMOs fully exploit the possibilities of new techniques in general, and use the options and potential of the WWW in particular, to disseminate these data and play an appropriate role in the National Geospatial Data Infrastructure?

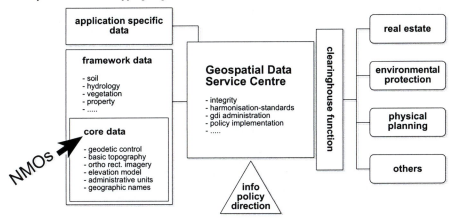

Figure 8.1 Position of NMOs in the Geospatial Data Infrastructure (after Groot and Kraak, 1999).

8.2 WHY WEB MAPS FOR NMOs?

From the foregoing, it is clear that National Mapping Organisations should explore their current use of the World Wide Web and start to think of their future use, if they have not done so already. The use of the World Wide Web as a first information source is already apparent for many applications, and there is no reason to assume that this is not the case for applications that need geographic base data. A very important and fast growing application area for NMO-based web maps can be found at websites that integrate information from different sources (Stähler, 1999). Organisations may want to offer web maps that guide customers to the closest branch office or another object (see for instance real estate websites like *URL 8.3*, that offer base maps as an index to locate areas of interest). These web maps may be based on NMO products or even derived from an online NMO service.

Additionally, the World Wide Web will become an important source of new NMO customers who may not yet be aware of what the NMO produces. Therefore, a website may be regarded as a means of advertising a NMO, in order to attract attention. But advertisement alone is not enough, an appropriate website should offer a well-balanced mixture of information, advertisement and entertainment (Figure 8.2). In the past, paper maps were the best means of advertising a NMO; nowadays, static and dynamic web maps can take over this role. An interested visitor to a NMO website is in search of information and expects to find at least some description of the typical NMO products or services on offer, which serves both information and advertisement purposes (Figure 8.3a): after finding the proper information, a visitor may become a customer.

However, it is not so easy to organise all information in such a way that suits visitors with different backgrounds, which may cause some considerable effort to locate the required information. An important service for less experienced visitors could be to offer instructive or even educational information on the cartographic discipline (Figure 8.3c), which may help in determining the NMO product most suitable to a visitor's query – this will serve both informative and entertaining tasks. In order to retain the visitor's attention, the entertaining contents of the website should certainly not be underestimated: samples of specific NMO tasks or topographic name games (Figure 8.3b) may seem trivial, but they may help specify the visitor's problems and demonstrate the NMO's expertise with new technology – a combination of entertaining and advertising functionality.

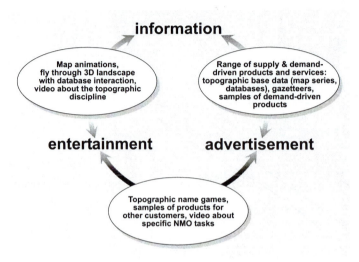

Figure 8.2 The three fundamental components for a well balanced website – advertisement, information and entertainment – with examples for NMOs.

Summarising, some examples of combinations between any two of the components of Figure 8.2 can be:

- *information and advertisement*: by providing an overview of products and services on offer, both supply and demand driven, accompanied by freely downloadable samples.
- *information and entertainment*: educational material, a fly-by animation through a virtual 3D landscape with database interaction or video fragments showing specific themes from the topographic discipline.
- *entertainment and advertisement*: a topographic name game, entertaining samples of products for other customers, video fragments about specific NMO tasks or organisation units.

A well considered balance among these components is critical, if one would want a visitor to return to the website again. Without an optimal balance the attention of the audience will be distracted more than was intended, which will lead to a rapid decrease of interest. The evaluation of this aspect is rather subjective, since it is largely dependent upon the scope of the site and the specific area of the discipline. For instance, a site containing detailed information on maps can be appealing for cartographers but at the same time may not be interesting for other disciplines. The applications mentioned and illustrated in Figures 8.2 and 8.3 are just some examples from the extensive list of possibilities for NMO web maps. From these, the most important will be treated in the following sections.

Figure 8.3a Information and advertisement – a sample of an analogue product of the Swiss NMO (reproduced with permission of the Swiss Federal Office of Topography (BA002035)) (*URL 8.4*).

Figure 8.3b Entertainment and advertisement – a puzzle on the website of the Finnish NMO (*URL 8.5*).

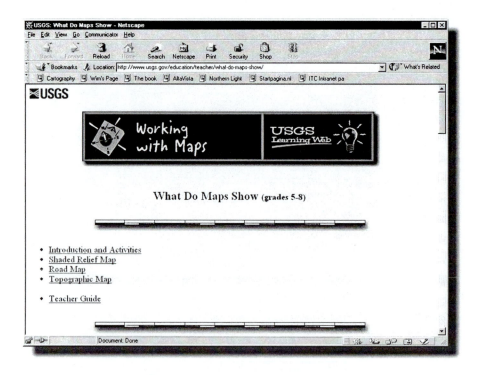

Figure 8.3c Information and entertainment – instructions on the USGS Learning website (*URL 8.6*).

8.2.1 Index to map products and services

Web maps can act as a very effective index to the standard NMO product line of both paper and digital map products. For their analogue map series most NMO websites use static, view only index maps that show the layout of map sheets over a country's area, for example as presented on the Northern Irish NMO website (Figure 8.4).

With simple alterations these maps can be transformed into static, interactive web maps. Clicking on or moving the cursor over a specific map sheet area will reveal its specifications in text format (e.g. the Netherlands Topographic Service, *URL 8.8*; or the Swiss NMO, *URL 8.9*; see also Chapter 7), possibly in combination with a map sheet extract presented as a static, view only web map (e.g. the Ordnance Survey of Great Britain, *URL 8.10*).

For the digital product line web maps can also serve as an index, although in this case the area is not necessarily subdivided into separate map sheets, since digital data can seamlessly cover a country's area. However, many NMOs offer their digital data in partitioned files that most often resemble the analogue map sheet subdivision. This relict of the past organisation of the production line restricts a map window to paper borders and will very likely disappear with the introduction of already available database technology that allows specifying a user-defined map window on a country's digital database

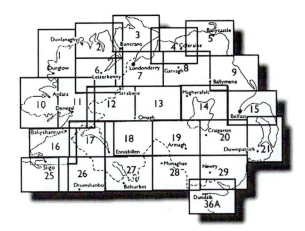

Figure 8.4 A web map as index to the analogue map series – Ordnance Survey of Northern Ireland (© Crown copyright, Permit ID1496) (*URL 8.7*).

A special service for which web maps are greatly suited is the presentation of an interactive gazetteer on a NMO website. A gazetteer is a list of all geographic objects that appear in the NMO maps and data sets, with their location coordinates. An interactive gazetteer on the WWW allows a search for geographic names and can return the result by means of a map extract that contains the corresponding object. In practice, the map samples presented are static, view only (e.g. the Australian NMO, *URL 8.11*); the "Get-a-map" service of the British Ordnance Survey (*URL 8.12*) offers a possibility to view map extracts at different scales – corresponding to the analogue map series – containing the requested geographic object. The Tiger Mapping Service of the US Census Bureau (*URL 8.13*) also allows searching the gazetteer via static, interactive web maps: clicking on objects may reveal information on coordinates and other typical census attributes.

8.2.2 Access to digital base data sets

It has already been stated that the Web will act as the backbone for the Geospatial Data Infrastructure: geodata are maintained by different mapping organisations and are therefore not necessarily stored at one location. Many countries have already established organisations that direct the developments and accessibility of GDIs or clearinghouses over the Web: web maps can provide or act as the links to the related mapping organisations or databases. For example, the US Federal Geographic Data Committee (*URL 8.14*) provides a (static, interactive) web map showing the *gateways* to the US-GDI.

Of all participating organisations, NMOs form an indispensable node in the potentially extensive network, being the providers of the core data. As the delivery of these data in digital form over the Web can be regarded as an important NMO service, a NMO website should provide the proper tools and information to access the data. The actual download of data may be restricted as a result of cost or copyright policies (see also Section 4.3). Whereas in North America geographic data are considered public domain and are free for download, NMOs in other parts of the world will charge at least for cost recovery. In those cases, a download facility is only possible via registered access. After logging in, the amount of downloaded data can be recorded, for which costs can be charged by sending an invoice to the customer – this method is, for instance, in use by the Finnish NMO (*URL 8.15*).

Another possibility is to provide at least the descriptive information of the digital data sets, which is often contained in metadata catalogues, and additional information on how to order the required data sets online from the appropriate organisation. Examples of this approach are the clearinghouse websites of Denmark (*URL 8.16*) and Uruguay (*URL 8.17*): both sites use static, view only web maps as samples and static, interactive web maps as the search facility for data in specific regions of the country. Sometimes it is also possible to download a sample data set of both digital raster and vector data that allows for experimentation by the customer to evaluate the data suitability (e.g. the Netherlands, *URL 8.18*).

Next to the typical supply-driven products, attention should be paid to the increasing demand for customer-defined services. Visitors may wish to order online digital base data sets that cover a user-specified area and only contain those data layers in which the user is interested. Of course, the site should contain some clear-cut samples of demand-driven products or services that the NMO has already carried out for other customers or can be achieved with NMO data sets, showing the NMO's capabilities.

8.3 CURRENT NMO ACTIVITIES ON THE WEB

8.3.1 Presence on the Web

With the above information on the potentials of the World Wide Web as a new instrument for map makers in mind, it is time to start an exploratory inventory of activities of National Mapping Organisations on the WWW. Such an inventory can begin with examining whether NMOs maintain a homepage on the WWW at all. For example, an overview of available homepages for NMOs worldwide is given in Figure 8.5; the corresponding links are available via *URL 8.2*. Currently, many NMOs in Europe, America, Southeast Asia and Oceania appear to be active on the WWW to some extent. Obviously, this map resembles the pattern of the global distribution of Internet users (*see* Figure 1.1): maintaining a website seems worthwhile for a NMO only if a sufficient number of potential Internet users exists in a country.

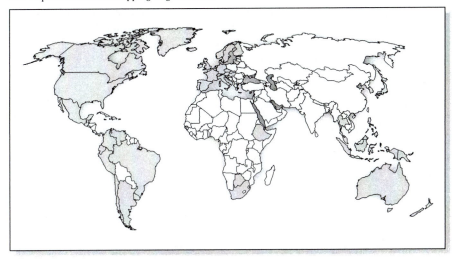

Figure 8.5 National Mapping Organisations on the Web (*URL 8.2*).

8.3.2 Inventory of current NMO website contents

A second step in the inventory of NMO activities on the WWW is the evaluation of the information content of their homepages. For this purpose a set of evaluation criteria can help assess the value of the homepages. Before defining evaluation criteria one should determine what kind of information is expected from a NMO, for example:

- Should the site not only list the supply-driven products but also demonstrate the capabilities to support demand-driven products (cf. Morrison, 1997)?
- Should the site contain map examples or complete map products?
- What should be the level of information detail: simple or extensive descriptions?
- What should be the technical level of applications that are presented: just simple text, functional graphics, or even databases that can be consulted with the browser via a map interface?

Table 8.1 lists some of the evaluation criteria that categorise the cartography-related key aspects (and could be applied also to any website discussed in the following application oriented chapters):

- *organisational information*: may cover the cartographic production line of the NMO as well as the organisation of the website (by means of a site map, see *URL 8.19*; see also Section 8.2.2).
- *analogue products information*: does the site contain descriptive information, (downloadable) sample maps, index maps (view only or interactive), price information? Is online ordering possible?
- *digital products information*: does the site contain descriptive information on digital data sets, downloadable data samples, price information? Is online ordering possible?

- *costs and copyrights* related to downloading information: are (subscription) costs reasonable and justifiable? What kind of restrictions are applied, e.g. watermarking of web maps? The site should offer clarity in complying with restrictions for dissemination of the information.
- *additional services*: if a NMO offers other services, such as customisation of map or data products, the site should contain related information and examples of customised results.
- *links*: the site should at least present links to the other organisations that use or process NMO data, or to the National Geospatial Data Infrastructure.

Table 8.1 A sample of a basic inventory of some NMO websites (for links to most NMOs on the WWW, see the ITC-website: *URL 8.2*).

Country	language¹	general info	index map²	Analogue maps	Analogue samples	Analogue price info	Digital maps	Digital samples	Digital price info	other services	links
Hungary	h	•									
Luxembourg	f	•	•	•	•	•				•	•
Netherlands	d/e	•	c	•	•	•	•	•	•	•	•
Norway	n	•	c	•	•		•			•	•
United Kingdom	e	•	c	•	•	•	•	•	•	•	•
Canada	e/f	•	c	•	•	•	•	•	•	•	•
USA - USGS	e	•	c	•	•	•	•	•	•	•	•
Argentina	e/s	•	•	•						•	
Brazil	e/p	•		•						•	•
Colombia	s	•	c	•			•			•	
Ecuador	s	•	•	•						•	
Mexico	e/s	•		•			•	•		•	
Botswana	e	•									
South Africa	e	•		•	•	•	•			•	•
Indonesia	i	•		•		•	•			•	•
Japan	e/j	•		•	•	•	•			•	•
Jordan	a/e	•		•	•						
Qatar	a/e	•		•	•	•	•			•	
Australia	e	•	c	•	•	•	•	•	•	•	•
New Zealand	e	•	c	•	•		•			•	•

Generally: • present
¹language: **a** Arabic; **d** Dutch; **e** English; **f** French; **h** Hungarian;
            i Indonesian; **j** Japanese; **p** Portuguese; **s** Spanish.
²index map: • present, but static and view only;
            c clickable map (static, interactive).

Table 8.1 only lists the presence or absence of these aspects, so does not indicate anything yet about the quality, functionality or extent if an aspect is present, neither does absence of the listed aspects directly imply that the website is of less value. When browsing through websites of different NMOs, it is obvious that the information content of each of these aspects is highly variable in quality and quantity.

Additionally, the table also lists the language which is used for the website: for example, the NMOs of Hungary and Norway only use their national language, which may give rise to the conclusion that these organisations strongly focus on their national audience. If international access is desirable, it is recommended to offer an alternative version in an international, easily accessible language. This version does not necessarily have to be as extensive as the original version as long as some general information offering contact possibilities is available

A considerable number of NMOs are already experimenting with online ordering facilities (mostly focused on their analogue product line): see, for example, the French IGN (*URL 8.20*) or the Danish KMS (*URL 8.21*).

From the websites explored for the purpose of making the table (other websites can also be accessed via *URL 8.2* where the table is also kept up to date) it is obvious that the web maps in use are all without exception either static, view only or static, interactive. Dynamic maps are hardly present – and if they are, then they serve demonstration purposes (*URL 8.22*).

8.4 RECOMMENDATIONS

From this chapter it should be evident that NMOs need to evaluate their current WWW activities, if they have not done so already. The most prominent arguments for this are:

- the WWW will play a prominent role in the national (global) geospatial data infrastructure: networking distributed data sets;
- the WWW can act as an advertising medium for NMO products and services (many people retrieve their first information on a subject from the WWW);
- the WWW can help explore new products and possibilities for NMO data sets.

The presentation of web maps forms the key instrument in providing an appropriate interface to these NMO tasks. In this respect the design of suitable web maps is crucial (see Chapter 7): index maps can be characterised by minimal information content (avoiding redundant details), whereas map samples need to reflect the full map content. The following sections sum up the considerations and recommendations that may support the determination of NMO activities on the Web. It is emphasised that these activities need not be limited to NMO websites alone NMO web maps can be plugged in to other websites as well, acting as an online location or route planning service (Stähler, 1999; e.g. *URL 8.22*).

8.4.1 NMO website considerations

There are already many guidelines available for designing good websites (*URL 8.23*). These guidelines are based on criteria that can be matched to the purpose of the website. The website of a NMO should meet the general criteria for good design as well as additional criteria of a cartographic nature. The nature of the anticipated audience should be carefully considered. An investigation may lead to the establishment of a common denominator for most NMO homepages. It is obvious that on one hand the intended audience is expected to be interested purely in the information that the NMO has on offer. On the other hand, the NMO will see its audience as potential customers, which indicates that the site will also have to act as some form of advertisement for its products and services. Aiming at an optimal balance between information and advertisement, the importance of the entertainment level of the website should not be underestimated: the website should attract and hold the attention of the intended audience. Entertaining action on the website could range from map animations or video fragments showing the work of a topographer in the field or showing the processing of data in a GIS, to a game which tests the user's topographic knowledge.

8.4.2 Minimal requirements of a NMO website

The NMO website should at least provide a presentation of the supply-driven products that are on sale: this could cover both the analogue map series and the digital base data sets that are developed, next to the availability of gazetteers. Naturally, this should not only be limited to simply providing names or numbers of map series or databases, but also provide clear insight in the precise specifications of the products on offer, from which the interested website visitor can deduce the product's fitness for use in relation to the intended application. Ideally, the website also contains some samples of digital raster graphics (scans of map fragments) or digital line graphics (a data set which is directly applicable in a GIS) with which a potential customer can experiment to evaluate the suitability of the product's format. Still, such a website would possess a mainly producer-oriented (read: supply-driven) character.

8.4.3 Towards a customer-oriented NMO website

To achieve a more customer-oriented approach, the website should also pay attention to demand-driven products. What kind of products can be delivered on demand, matching specific user requirements? There should be an option for online ordering of printed maps on demand for a user-specific map window, that may cover an area of neighbouring map sheets, and which contains only those map features that the user requires. Such an option could as easily be extended to the ordering of the corresponding digital base data sets. Next to such a range of products, the website can pay attention to the services which the NMO is capable of delivering on demand or cross-refer to other institutes or companies that provide such services making use of NMO data sets. Of course, the site should contain some clear-cut samples of demand-driven products or services that the NMO has already carried out for other customers or that can be achieved with NMO data sets, showing the NMO's capabilities.

Next to presenting current supply-driven and demand-driven products or services, the ideal website should also pay attention to possible products that may be in wide use in the near future, in a way that attracts potential customers who are not yet fully aware of what NMO data have to offer. A demonstration of such an advanced visualisation could be a flyby through a virtual reality environment providing user interaction with the underlying data sets. At present such a sample will most likely perform an entertaining factor on the website for most visitors, but at the same time it is of special interest for researchers who are working on virtual 3D environments and who can use the digital topographic base data as the framework of their information systems.

URLs

URL 8.1   The US Federal Geographic Data Committee <http://www.fgdc.gov>

URL 8.2   ITC switchboard to all known NMO websites world wide
            <http://www.itc.nl/~carto/nmo>

URL 8.3   Homenet - online real estate service <http://www.homenet.com>

URL 8.4   Swiss Federal Office of Topography – illustration of analogue map
            products <http://www.swisstopo.ch/de/maps/lk/500er.htm>

URL 8.5   The Finnish National Land Survey <http://www.nls.fi/index_e.html>

URL 8.6   The USGS Learning website
            <http://www.usgs.gov/education/teacher/what-do-maps-show/index.html>

URL 8.7   Ordnance Survey of Northern Ireland – index map 1:50 000
            Discoverer Series <http://www.doeni.gov.uk/ordnance/catalog/34.htm

URL 8.8   Topographic Service of the Netherlands – clickable index maps
            <http://www.tdn.nl/uk/topokrt1.htm>

URL 8.9   Swiss Federal Office of Topography
            <http://www.swisstopo.ch/en/maps/lk/25over.htm>

URL 8.10 Ordnance Survey of Great Britain
            <http://www.ordsvy.gov.uk/products/Landranger/lrmsearch.cfm>

URL 8.11 Australian Land Information Group
            <http://www.auslig.gov.au/mapping/names/names.htm>

URL 8.12 Ordnance Survey of Great Britain "Get-a-map" service
            <http://www.ordsvy.gov.uk/getamap/index.html>

URL 8.13 US Census Bureau Tiger Mapping Service
            <http://www.census.gov/main/www/srchtool.html>

URL 8.14 US Federal Geographic Data Committee <http://fgdclearhs.er.usgs.gov/>

URL 8.15 Finnish National Land Survey data download facility
            <http://www.kartta.nls.fi/karttapaikka/eng/services/ammattilaisen.html>

URL 8.16 Clearinghouse of Denmark
            <http://www.kms.dk/geodata/topografiske/index_en.html>

URL 8.17 Clearinghouse of Uruguay <http://www.clearinghouse.com.uy/>

URL 8.18 Topographic Service of the Netherlands
            <http://www.tdn.nl/uk/digindex.htm>

URL 8.19 Austrian Bundesamt für Eich und Vermessungswesen
            <http://www.bev.gv.at/graphische_uebersicht.htm>

URL 8.20 IGN France <http://www.ign.fr/GP/cartes/bleue.html>

URL 8.21 KMS Denmark <http://www.kms.dk/kortbutik/bestilling_en.html>

URL 8.22 Ordnance Survey of Great Britain
            <http://www.ordsvy.gov.uk/samples/gislab.html>

URL 8.23 Information quality WWW virtual library
            <http://www.ciolek.com/WWWVL-InfoQuality.html>

REFERENCES

Groot, R. and Kraak, M.J., 1999, Challenges and opportunities for national
  mapping agencies development of national geospatial data infrastructure
  (NGDI). In *Proceedings UN-CODI Conference (E/ECA/DISD/CDI.1/38)*, Addis
  Abeba, Ethiopia, (Addis Abeba: Committee on Development Information)
  pp. 1-11.
McLaughlin, J. and Coleman, D., 1999, Geomatics at the end of the century:
  framing a new agenda. In *Cambridge Conference Papers*, (Southampton:
  Ordnance Survey), paper 1.1.
Morrison, J.L., 1997, Topographic mapping in the twenty-first century. In
  Framework for the world, edited by Rhind, D., (Cambridge: Geoinformation
  International), pp. 14-27.
Peuquet, D.J. and Bacastow, T., 1991, Organisational issues in the development of
  geographical information systems: a case study of U.S. Army topographic
  information automation. *International Journal of GIS*, **5**, (3), pp. 303-319.
Rhind, D. (ed.), 1997, *Framework for the world* (Cambridge: Geoinformation
  International).
Stähler, P., 1999, Potentiale des vernetzten Medienmanagements. In *Proceedings
  Web.mapping.99*, (Karlsruhe, Fachhochschule) pp. X.1-X.14.

CHAPTER NINE

Web maps and tourists

Allan Brown

9.1 TOURIST INFORMATION ON THE WEB

Tourism is of such enormous importance to many economies, including those of highly developed countries, that it is not surprising that there is a very large and rapidly growing number of websites devoted to it. A vacation trip is not one of those items that one can see and touch before purchase. Choosing and planning a vacation have to be based on information supplied by an intermediary (e.g. travel agent or friend) or directly by the provider. To make travel and accommodation reservations one normally goes to a travel agent. Among the great advantages of the WWW in the tourism business is that it enables the customer to have much easier direct access to a very large amount of up-to-date information and to do the reservations from home. Many government-financed tourist offices have recognised the importance of the Web and have put much effort into developing good sites.

In most cases (at present at least, although due to technological developments this is about to change) the tourist has no access to the WWW while on a vacation trip. Therefore he or she has to try to collect information before departure. While on a trip, conventional guidebooks and maps will still be used, so it is not surprising that one subsidiary aim of many tourist sites is to sell these. The Web perhaps does not (yet) compete with guidebooks in portability, but it can offer things which guidebooks cannot, e.g. sound and video sequences, virtual reality models (still rather limited), direct links to many other sources of information and online reservation services. A full guidebook may in fact be put on the Web (paid for by advertising revenue), or the tourist can compile a personalised guidebook before departure by simply printing the selected web pages. The aim of the makers of tourist web pages is not only to provide information but also actively to promote the product, which must therefore be made to appear as attractive as possible.

9.2 EXAMPLES OF OFFICIAL AND COMMERCIAL WEBSITES

Maps can play a very important part in providing information to tourists, so it is rather disappointing to find that maps are not used as much as they could be on tourist websites, and where they are their design and use is often not very imaginative. An example is the award-winning official site of the Tourist Office of Spain (*URL 9.1*), containing (in summer 2000) only a few small, simple, non-clickable maps. This may perhaps reflect the particular market. Countries such as Spain which rely very heavily on packaged holidays to beach resorts may feel less need for maps on their websites than those which rely more on individuals or

families travelling around by car or public transport. Ireland is an example of the
latter situation.

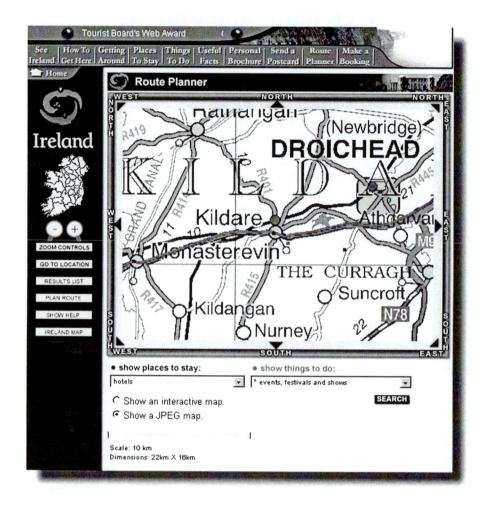

Figure 9.1 An interactive map from the Irish Tourist Board website
(courtesy of Bord Fáilte/Irish Tourist Board) (*URL 9.2*).

Despite the nature of tourism to the island, The Irish Tourist Board site (*URL
9.2*), also an award winner, only introduced the interactive use of maps, based on
maps of the Ordnance Survey of Ireland, in late 1999. This website can be taken as
a general illustration of the importance of the Internet to tourism and (when
accessed in December 1999) it also illustrates many of the specific capabilities of
tourist maps on the WWW. The Board stated in mid-1999 that so far that year
there had been an increase of 68% in the number of visitors to the site compared to
1998, when the total was 1M. The Internet is now the most important source of
information for visitors to the country. The site offers "over 11 000 places to stay

and 10 000 things to see and do". It contains information in various categories, including how to get there. There is an online booking service to over 3000 accommodation addresses. Visitors to the site can easily compile and print their own personal brochure.

On the cartographic side, there are maps of the country available at four zoom levels, so following a typical stepped zooming system (Section 7.2). Judging by their design, all the maps appear to be scanned paper maps, with their typical limitations. Since the provider has no control over the viewing situation, the scales of these maps are not given as representative fractions but as bar scales (as recommended in Chapter 7). Very approximately, the maps appear on the monitor as 1:5 000 000, 1:1 250 000, 1:200 000 and 1:60 000 scale, with in addition a large scale (1:4000) map of the city of Dublin. Except for the small scale general map, all the maps are shown on the screen as small windows following a tiling system. To move the window to an adjacent tile one has to click on one of eight arrows at the sides and corners (see Figure 9.1).

To plan a route one can enter up to 10 places. The route connecting these then appears in a distinctive purple colour on a general map of the country. The chosen route appears at all zoom levels. A written turn-by-turn description of the route is also supplied. Under each map is a drop-down list of categories of places to stay and things to do in the area covered by the map. Using a free plugin, a selected category from each list appears as small symbols in the map window. Putting the mouse over the symbols reveals their names and double clicking takes one to a page of information.

This Irish website uses maps supplied by the official government mapping organisation. A very common situation on tourist websites, however, is that maps are downloaded from other sites. An example of a source that is frequently used is the website containing scanned paper maps provided by the University of Texas (*URL 9.3*). Many of the cartographically more sophisticated sites use maps supplied by commercial third parties such as MapQuest, Map Blast! or Multi Media Mapping, which may themselves get their data from other geodata suppliers. These companies offer general mapping, using specially prepared maps, with the possibility to zoom in or out along a stepped range of perhaps ten scales, and to pan tilewise. Panning is necessary since the limited resolution and size of the screen prevents a large map from being shown all at once. Furthermore, the use of small windows decreases download time. These maps are often static, view only. On MapQuest's own site (*URL 9.4*) it is however possible to personalise a map of a city by adding symbols for hotels, restaurants, attractions, etc. Figure 9.2 shows a page from the site, showing a part of New York with attractions symbolised. The selected map can be customised to a certain extent, e.g. made larger, different background colour. Once the user is satisfied, the map can be printed or downloaded to a handheld device.

9.3 MORE ADVANCED CARTOGRAPHIC TECHNIQUES

Clickable maps are used on many tourist sites. Sometimes clicking is used to go to a larger scale map, more often it is used to provide textual information, photographs (including 360° views), Virtual Reality models, booking forms for

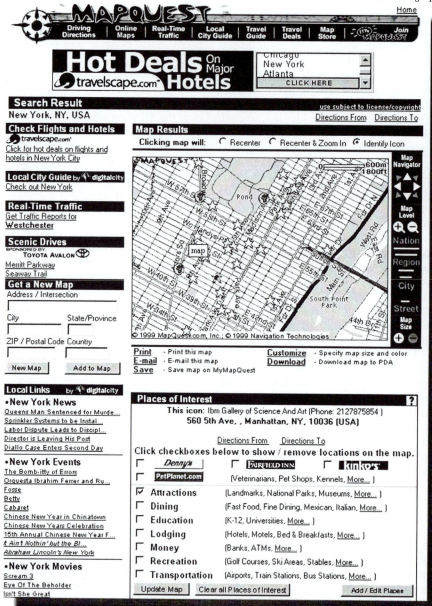

Figure 9.2 **An example of a MapQuest page (*URL 9.4*).**

accommodation, etc. A rather advanced example of this principle is provided by the Channels site for Amsterdam (*URL 9.5*). A red spot on the map shows one's position, with street photographs taken from that spot also on display. One can click on a direction arrow to move to a nearby spot. To move to a different part of the city one has to "travel" by clicking on the appropriate tramline number or by taking a "taxi ride".

As has been mentioned already, maps on websites are very often displayed in small windows to reduce downloading time. This necessitates being able to move or "pan" the window to other parts of the map. A disadvantage of the standard tile-and-arrow method, as used for example on the Irish Tourist Board and MapQuest sites, is that the adjacent map tile has to be regenerated each time. MapQuest also offers the possibility to recentre a map by a small distance by clicking on the new centre desired but here again the map has to be regenerated. "Dragging" offers a quicker alternative. This technique is used by the WorldWeb Technologies TravelGuide to British Columbia (*URL 9.6*). They have also made much use of the "mouse-over" technique (see Figure 9.3). The original maps used are well-designed maps supplied by another company, also available as printed copies.

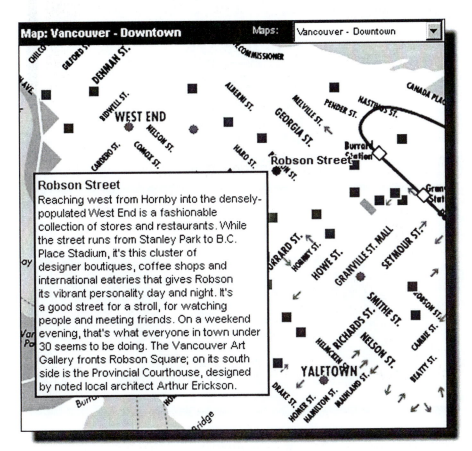

Figure 9.3 "Mouse-over" on the WorldWeb TravelGuide to British Columbia (*URL 9.6*).

The designers of some tourist information sites have paid special attention to the technical aspects of downloading maps. An example is a site offering scanned historical maps of Paris, in which the user can vary the scale and size of the tile required and its resolution, in order to optimise search and download time (*URL 9.7* and Figure 9.4).

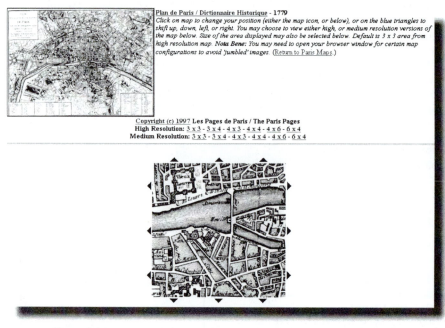

Figure 9.4 Selection of size and resolution for historical maps of Paris (*URL 9.7*).

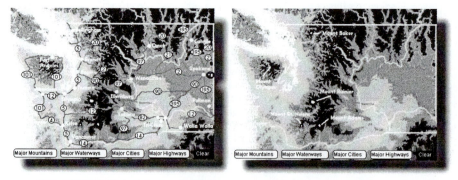

Figure 9.5 Variations of the same map using a "control-panel" legend, courtesy of Travel in Washington (*URL 9.8*).

Some of the ways in which the clickable map technique can be used in the context of tourist mapping on the Web have already been mentioned. Another example of its use (*URL 9.8* and Figure 9.5) is to allow the user to click on legend items to cause them to appear on the map, an example of the control-panel legend mentioned in Chapter 7. The application illustrated in Figure 9.5 was implemented in Shockwave format (SWF, see Appendix A).

9.4 DOWNLOADING AND PRINTING HIGH QUALITY TOURIST MAPS

It is always of course possible simply to print a web page directly, on a monochrome or colour printer. The printing option offered on most tourist sites is actually little more than this, perhaps with some elementary layouting. The actual quality of the print therefore leaves a lot to be desired. If a designer would like the user to be able to produce a higher quality print, an obvious option is to make a high resolution GIF or JPEG image available. If the resolution is too high, however, the map will take a long time to download. A somewhat lower resolution will produce an acceptable print at a small scale. Since most users have only an A4 (or perhaps A3) printer at their disposal, it is reasonable to specify the lowest resolution on the web map which will still give a good result when printed at A4 or A3 format. However, if a part of the map is enlarged, the result is a "blocky" or unsharp image. Figure 9.6 illustrates this. It is an inset from a JPEG map of Boston (*URL 9.9*).

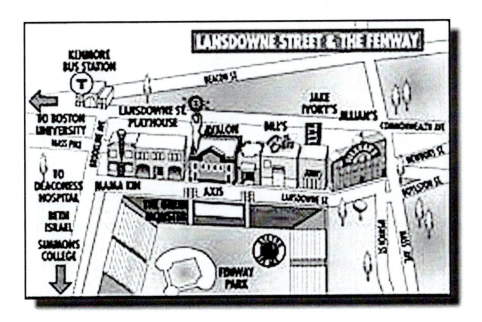

Figure 9.6 A JPEG file printed much enlarged to show the "blocky" image (*URL 9.9*).

A different approach is to provide maps on the website in high-resolution PDF format. These can be printed at almost any size and still give good results.

Provided they were not security-protected by the provider, they can also be imported into other software packages and edited if desired. The example illustrated in Figure 9.7 is taken from a California site (*URL 9.10*). The State is divided into regions, with overviews provided as fast-loading low-resolution JPEG images. Clicking on the map or the name of the region of choice downloads a PDF map. One can pan to a different part of the map by using the slide bars or a "dragging hand". One can also zoom in and out in fixed steps, with the advantage that even at the very largest scale the image is sharp. A search facility is provided – typing in the name of a place causes the window to pan to the area and the place to be highlighted. The map itself is not clickable. The PDF map should print out on any printer (but it was found that the colours are very different on different printers). The PDF file can be imported into a graphics package and edited. Figure 9.8 shows the result of processing the file in Macromedia FreeHand. In this case the main font used on the original map was not available, so a substitute font had to be chosen.

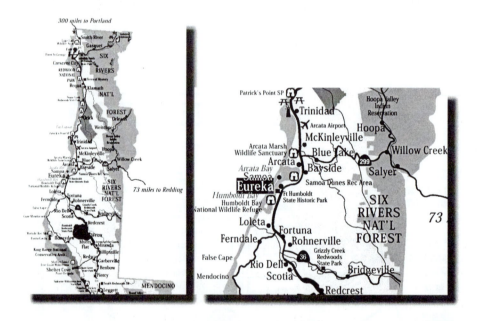

Figure 9.7 A PDF file of part of northern California, rasterised in Adobe Photoshop, at different scales (with the town of Eureka highlighted on the larger scale map) (*URL 9.10*).

Some sites actively encourage the user to download and print very high quality maps. This is particularly the case in the USA where government information is provided free of charge, except for the cost of the carrying medium. The National Park Service is a good example of this policy. In the case of the North Cascades National Park (*URL 9.11*), for instance, maps are provided in three different formats: PDF for screen viewing; Adobe Illustrator 6.0 for print production; high-resolution JPEG (very large and slow to download) of the hillshaded map, that can be printed via a graphics package. On the PDF version, searching for a place, panning and zooming are done as for the California maps mentioned above.

Figure 9.8 The PDF file of Figure 9.7 processed using Macromedia FreeHand, with manual font substition for unavailable fonts.

9.5 FUTURE PROSPECTS

There is no doubt that there is still a tremendous growth potential for the supply of tourist information over the Web. Commercial companies are already competing strongly with each other. More and more governments will realise that their official sites can play an important role in generating tourist business. Tourist websites will contain more information, they will become better designed and more efficient in terms of fast downloading and ease of navigation within the site. With the coupling in prospect of mobile telephone technology and Personal Digital Assistants, attention will need to be paid to supplying information specially designed for these users.

What is perhaps surprising at present is that, apart from the route planning sites, many tourist sites do not use maps at all, or where they do then very unimaginatively, e.g. a couple of paper maps scanned at low resolution. It is likely, however, that more and more tourist website designers will realise the value of maps on their sites. The map also forms a very efficient method of organising information on the Web. Many tourists will only be visiting a region of a country, so they require detailed information only of that region. The clickable map and "mouse-over" techniques are ideally suited for this purpose.

As maps become more common on tourist websites, it is probable that users will become more demanding and more critical. They will want clear, well-designed maps that will also look good when printed. The problem here is that if such maps are in one of the standard raster formats at high resolutions, then downloading could be very slow. As has been shown in the previous section, providing first a small file (e.g. in JPEG) format so that the user will know in advance what he is going to get, coupled with a high quality PDF map, is a solution. Panning, zooming and highlighting selected places are easy to organise on PDF maps and they can also be made clickable.

Although we can expect that more website designers will use PDF maps in the medium term, in the longer term providing maps in a vector format, e.g. SVG, may turn out to be more efficient (Appendix A). This of course would reduce the presently widespread use of scanned paper maps on tourist sites. It would also present a challenge to cartographers to produce good vector maps within the technical constraints of the Internet.

Some tourist websites already attempt to provide a "virtual tour". The Channels site for Amsterdam has been mentioned already (Section 9.3). Museums have also discovered the possibilities. On the official site of the Louvre (*URL 9.12*), for example, one can click on many different points on a plan of the building to get a panoramic VR image. This technology is also in use on some general tourist sites. For examples of panoramas in both QuickTime VR and Live Picture Zoomit formats, see *URL 9.13*.

VRML models are not as yet in very common use, but see *URL 9.14* for a model of Schiphol airport and *URL 9.15* for a model of the Expo 2000 exhibition area in Hannover. These models have the advantage of providing very realistic virtual tours, so it is possible that sites that already attempt virtual reality, such as museum sites, will adopt this technique. Another possible use of the technique could be in modelling ski resorts, many of which already commission panoramic views (*URL 9.16*). The advantages of a model over a scanned panorama are obvious: the potential client could view the slopes from any angle and even attempt a "virtual descent"!

Another development to be expected is the increasing use of sound. It could be very useful, for example, to hear exactly how a place name is pronounced when one clicks on it. Another application could be to hear in conjunction with a chosen route on a route planner that delays can be expected because of road works between certain dates, or even that traffic jams are being experienced at that very moment.

In conclusion it is safe to forecast that tourist website designers will increasingly be on the lookout for ways of making their sites and the maps on them as attractive and efficient as possible. They will try to make use of the latest technological developments to bring the information quickly and easily to the customer. Considering the economic importance of tourism, we can also assume that there will be sufficient funds available to support these aims.

URLs

URL 9.1 Tourist Office of Spain <http://www.tourspain.es/turespai/marcoi.htm/>
URL 9.2 Irish Tourist Board <http://www.ireland.travel.ie/>
URL 9.3 University of Texas Map Libraries
             <http://www.utexas.edu/depts/grg/virtdept/resources/map_libs/map_libs.htm>

URL 9.4 MapQuest <http://www.mapquest.com/>

URL 9.5 Channels <http://www.channels.nl/>

URL 9.6 WorldWeb TravelGuide to British Columbia
  <http://vancouver.discoverbc.com/>

URL 9.7 Paris Pages <http://www.paris.org/>

URL 9.8 Travel in Washington
  <http://www.travel-in-wa.com/travel/TOPO/topo.html>

URL 9.9 Historic Tours of America: Map of Boston
  <http://historictours.com/boston/bostonmap.htm>

URL 9.10 California interactive maps <http://gocalif.ca.gov/maps/>

URL 9.11 North Cascades National Park <http://www.nps.gov/carto/NOCA.html>

URL 9.12 The Louvre official site <http://www.louvre.fr/>

URL 9.13 Examples of VR and Zoomit <http://www.culture.com.au/virtual/>

URL 9.14 3D model of Schiphol airport
  <http://flightinfo.schiphol.nl/engine/index_def.html?lang=en&page_nr=590>

URL 9.15 3D model of the Expo 2000 exhibition area, Hannover
  <http://www.dtag.de/expo2000/terra/>

URL 9.16 Ski maps <http://www.skimaps.com/>

CHAPTER TEN

Web maps and atlases

Menno-Jan Kraak

10.1 ATLAS CONCEPTS

Atlases are probably seen as the best known and ultimate cartographic products. At home and school everyone has had to familiarise themselves with atlases. Their use is primarily to locate geographic phenomena such as places, rivers and regions, or to understand geospatial patterns related to the physical or socio-economic environment. Many armchair journeys have been made guided by an atlas. An atlas is defined as an intentional combination of maps, structured in such a way that given objectives are reached (Kraak and Ormeling, 1996). Ortelius' Theatrum Orbis Terrarum published in 1570 is recognised to be the first atlas according to this definition. In this context the word "atlas" was first used in 1585 by Mercator in his publication Atlas sive Cosmographicae Meditationes de Fabrica Mundi et Fabricati Figura. Over the years different types of atlases have developed.

Today one can distinguish among reference atlases, school atlases, topographic atlases, national atlases and thematic atlases. The first aim at giving access to as many geographic objects as possible. This tradition started during the 19th century in Germany with products like the Stieler Weltatlas. Today's major representative is the monumental Times World Atlas (1999). It is argued this might be one of the last paper atlases of this magnitude, especially when electronic atlases, such as Microsoft's Encarta, offer access to 8 times as many objects. The content of school atlases is much influenced by the geography curriculum in a country, but in general offers a generalised view of the world, with special attention to aspects of the home country. The Dutch Grote Bosatlas (1995) is a prime example. Full electronic versions of school atlases do not yet exist. Topographic atlases are often a book-form issue of a topographic map series. Some exist on CD-ROM. National atlases contain a comprehensive combination of geographical data representations that cover the country in a complementary way. As such they offer a contemporary view of geographic aspects of a country. Many countries have a national atlas and several are working on electronic or WWW versions. Thematic atlases are devoted to a single theme, such as environment, census or geology.

All paper atlases have some structure, which is designed to reach the atlas objectives. Their structure is expressed by the map scale, and the sequence of the maps. For instance, in school atlases most maps are devoted to the home country. These are also the maps with the largest scale. Usually, those are followed by maps of the continent where the country can be found. The continental maps are followed by an even smaller set of maps of the remainder of the world. The atlas structure will also define how well the user can execute comparison of geographic, thematic and temporal aspects of particular areas. Access to paper atlases can e obtained via the contents pages, map index, names index or by random browsing.

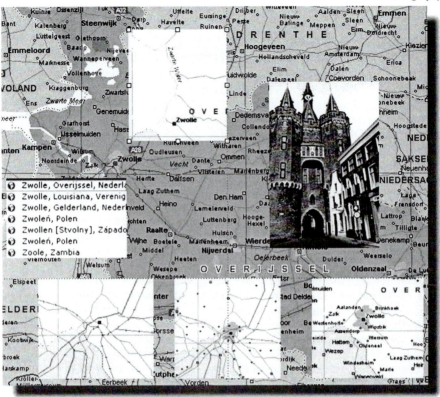

Figure 10.1 Views from an electronic multimedia atlas (examples from Microsoft's Encarta 2000©).

The development of electronic atlases started at the end of the eighties. The *Atlas of Arkansas* is seen as the first electronic atlas (Smith, 1989). It just holds a set of static maps that can be accessed via a menu. Interestingly enough the development of the electronic atlases has some parallels with the introduction of the computer in cartography. In the beginning the development was limited by factors such as storage size and screen resolution. The first atlases were just copies of paper editions, and are known as view only atlases. Later the producers started to use the additional options the digital environment offers: interaction and dynamics. This introduced interactive and analytical atlases. These atlases allow the user to decide on the map detail viewed and often also on its contents – Microsoft's *Encarta* (2000) is an example (see Figure 10.1). It offers an interactive relation between the map and other media via hyperlinks. Multimedia elements such as text, images and animations can be linked to the map. Some of these links give access to locations on the WWW. Analytical atlases should be seen as interactive atlases with (some) GIS functionality. Atlases are also found on the WWW. The types found are reference atlases, thematic atlases, and national atlases.

10.2 WEB ATLASES

Reference atlases on the Web are of different kinds. Some are simple and could be called clipart-collections (*URL 10.1*). Although they may offer, for instance, only outlines of countries they can still be useful for particular applications. Others offer maps on country level (*URLs 10.2 & 10.3*). Special cases are the country maps in the online Encyclopedia Britannica (*URL 10.4*). These clickable maps function as entry to the encyclopedia articles, when available, of the geographic elements on the maps. Some web atlases are derived from CD-ROM editions such as Encarta (*URL 10.5*). The most detailed ones allow the entering of queries up to street level (*URLs 10.6 & 10.7*). The response will depend on where on Earth the requested location is found. North America and Europe are very well covered because data used are derived from those data sets created for car navigation systems or made available by governmental mapping organisations. Some reference atlases have access restriction, and for more detail you have to pay. Most of the reference atlases allow one to input a geographic name and will return a map. Some allow for zoom and pan via a menu or are clickable. Based on the classification given in Figure 1.2 these are all static maps. Several publishers have put their paper atlas copies online. A prime example is the seventh edition of the National Geographic world atlas (*URL 10.8*). Although web functionality is added, it remains primarily a copy of the paper atlas, and one has to zoom to "paper atlas level of detail" to get a readable map on screen.

Thematic atlases on the WWW deal with topics such as history (*URL 10.9*), census (*URL 10.10*), and elections (*URL 10.11*). Again most of these atlases present view only maps. However some do present more advanced mapping represented by animation or interactive selection and classification.

Of all atlases the national atlas is the most special (Ormeling, 1979). Nations tend to consider the publication of a national atlas as a matter of self-respect, even of national pride. National atlases map all the particularities of a country, often starting with the physical geography followed by maps of all kinds of socio-economic topics. On the WWW only a few national atlases exist (see Figure 10.2). Well known are those from Canada (*URL 10.12*), the United States (*URL 10.13*) and Switzerland (*URL 10.14*). The first two are part of the national geospatial data infrastructure (Frappier and Williams, 1999; Palko, 1999). The last aims at innovative graphics (Neumann and Richard, 1999), but is still in an experimental phase. National atlases also exist in regional versions, often to contribute to a region's identity. An example is the Atlas of Quebec (*URL 10.15*).

By presenting a detailed view of the physical and social aspects of a country, the atlas as such is not only a mechanism to inform, but could be part of the geospatial WWW search engine – another entrance to the geospatial data organised via a clearinghouse. Geospatial data providers participating in a clearinghouse are also the data sources for a national atlas. The idea is that each organisation makes up-to-date data available at a certain level of aggregation. One can imagine that the National Mapping Organisation offers the data for a country's base map, while the other organisations offer data on the themes they deal with. Information at the national level is offered free of charge. More detailed information can be obtained from the respective organisations and should be accessible via links in the atlas and/or pointers at the clearinghouse.

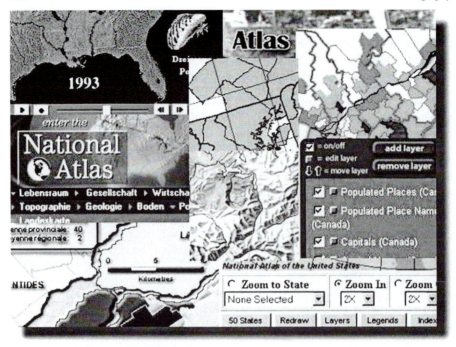

Figure 10.2 A mosaic of national atlases.

Based on a map design framework the maps could be directly generated by the participating organisations according to a National Atlas Standard that would guarantee up-to-date data and maps. Design rules could be part of, or derived from, the metadata needed for the clearinghouse, to avoid double effort in creating and maintaining atlas pages by the participating organisation. However, a simple design of static maps is not encouraging for most visitors to an online national atlas. They have a different attitude to maps compared to those reading a traditional paper atlas. Those new users are probably eager to click on the maps, and expect action. The national WWW atlases mentioned above have maps in all categories of the classification scheme found in Figure 1.2. However, it is clear the static maps are in a clear majority.

An interesting discussion involves the interface issues around web-atlases (see also Chapters 3 and 4). Web and electronic atlases require a specific interface regarding navigation and orientation. Examples are being able to trace the route followed through the atlas, showing where in the atlas the user is (e.g. inset overview map). These functions are closely linked to general web browser functionality. Specific cartographic aspects would involve being able to query the map for coordinates, to have legend functions available (see Chapter 7) and to be able to query a database behind the map. Other basic functions include being able to pan and zoom or even to rotate the maps, to create one's own maps (to choose the symbology or to select the data variable to be mapped), to change map projections and, typically related to atlas use, to be able to compare several maps in a single view.

The paper and electronic world each have their advantages while browsing an atlas. Paper atlases (as well as books) have the initial advantage of random browsing. During paper browsing your eye may fall on a particular geographic object such as a region or town, which trail you then try to follow. Starting digital browsing, your eye may never be in touch with this object at all, since often random browsing in an electronic or web atlas means following pre-cooked paths or typing keywords. However, as soon as you intend to follow a particular path the hyperlinks available to help you do this in an electronic or web atlas are not found in the paper atlas. These hyperlinks are a clear advantage, and although pre-cooked they are very helpful.

URLs

URL 10.1   Graphic Maps <http://www.graphicmaps.com/graphic_maps.htm>
URL 10.2   Magallan Geographix <http://www.maps.com/>
URL 10.3   CIA <http://www.odci.gov/cia/publications/factbook/index.html>
URL 10.4   Encyclopedia Britannica <http://www.britannica.com>
URL 10.5   Encarta <http://encarta.msn.com/maps/mapview.asp>
URL 10.6   MapQuest <http://www.mapquest.com/>
URL 10.7   MapBlast <http://www.mapblast.com/mblast/index.mb>
URL 10.8   National Geographic <http://www.nationalgeographic.com/mapmachine>
URL 10.9   Historical atlas of the twentieth century
            <http://users.erols.com/mwhite28/20centry.htm>
URL 10.10 Niagara Census < http://www.brocku.ca/maplibrary/atlas96/atlas96.shtml >
URL 10.11 Elections <http://www.wahlatlas.de/fr_java.html>
URL 10.12 National Atlas Canada < http://www.atlas.gc.ca/english/index.html>
URL 10.13 National Atlas USA <http://www.nationalatlas.gov/>
URL 10.14 National Atlas Switzerland <http://www.karto.ethz.ch/atlas/>
URL 10.15 Atlas de Quebec
            <http://www.unites.uqam.ca/atlasquebec/cadres/accueil.htm>

REFERENCES

Encarta Worldatlas, 1999, (Seattle: Microsoft).

Frappier, J. and Williams, D., 1999, An overview of the national atlas of Canada. In *Proceedings of the 19th International Cartographic Conference ICC99, Ottawa*, (Ottawa: Canadian Institute of Geomatics), pp. 261-266.

Grote Bosatlas, 1995, (Groningen: Wolters Noordhoff).

Kraak, M.-J. and Ormeling, F. J., 1996, *Cartography, visualization of spatial data*, (London: Addison Wesley Longman).

Neumann, A. and Richard, D., 1999, Internet atlas of Switzerland new developments and improvements. In *Proceedings of the 19th International Cartographic Conference ICC99, Ottawa*, (Ottawa: Canadian Institute of Geomatics), pp. 251-260.

Ormeling, F. J., 1979, The purpose and use of national atlases. *Cartographica,* **16,** (Monograph 23), pp. 12-23.

Palko, S., 1999, Partnership and the evolution of the national atlas of Canada. In *Proceedings of the 19th International Cartographic Conference ICC99, Ottawa,*

(Ottawa: Canadian Institute of Geomatics), pp. 275-284.

Smith, R. M., 1989, *Atlas of Arkansas.* (Fayetteville: University of Arkansas Press).

Times Atlas of the World: Comprehensive Edition, 1999, (London: Times Books).

CHAPTER ELEVEN

Web maps and weather

Connie Blok

11.1 INTRODUCTION

The term "weather maps" in this chapter loosely refers to any cartographic depiction of weather or weather-related phenomena, including climatic maps. A map showing temperature and precipitation, the UV-index, an enhanced cloud cover image and an animation of the dynamics of the ozone layer all fit into this category. Although there is a broad range of cartographic weather depictions, most of them can be related to two main goals of meteorologists and other atmospheric scientists: diagnosis and/or prognosis. Weather maps that deal with diagnosis describe current and past conditions of the atmosphere, while the prognosis maps attempt to project past and current conditions and trends forward in time.

Meteorology is practised in a map-rich environment. Maps play an important role in the processing of weather data and in the dissemination of information. Huge numbers of maps are produced. In fact, there is no other activity in which such a large quantity of *new* data is mapped on a daily basis (Carter, 1998). Way-finding and tourist mapping may nowadays seem to take the lead because sites like MapQuest, (*URL 4.8*) generate enormous numbers of maps (see Section 4.2). However, although users may add their own "points of interest" these maps predominantly consist of user-defined selections from already existing databases.

It certainly makes sense to represent weather data cartographically, because the spatial and temporal contexts are essential for the interpretation of weather characteristics. The value of the spatial context for atmospheric scientists was already demonstrated by the first weather maps in the early 19th century. They confirmed the hypothesis of circular and converging winds around a low pressure centre. More recently, measurements revealed variations in the amount of ozone in the stratosphere, but the "hole" in the ozone layer was only noticed after visualisation of the data (Monmonier, 1999). Meteorologists deal with large amounts of location-specific data, which can only be comprehended in a spatial context. In addition, the temporal context is required to determine which trends there are and to forecast, for example, speed and direction of movement of fronts and the strength of storms and showers (see Figure 11.1). Last but not least, many people without expert knowledge, even laymen, benefit from the spatial and temporal contexts offered by weather maps.

Because of the time-sensitivity of the information, the Internet is a suitable medium to disseminate weather maps as it allows near real-time display (see Chapter 1). The Web also offers opportunities to disseminate (at low costs) a much broader range of maps and other weather graphics than any of the conventional mass media. Someone searching the Web will soon discover that it is an almost inexhaustible source to study aspects of what Monmonier (1997) has called the "geography of the atmosphere". In most cases one can access maps (or other information) generated from processed data. This makes sense, because expert knowledge is required to interpret and model the many parameters of the weather, a phenomenon that is continuous in space, extends in three dimensions, and is continuously changing. However, several (mainly North American) sites also enable the user to download data, e.g. NASA's space flight centre Goddard DAAC for satellite data (*URL 11.1*) and NOAA's National Climatic Data Center (NCDC) for satellite, radar and climate data (*URL 11.2*). NCDC also provides an online tool to visualise the data available (see also Sections 11.3.4 and 11.4 below; *URL 11.3*); other examples are UM Weather with a variety of weather-related software (*URL 11.4*) and Meteo Consult, the Netherlands, with "Weerbeeld", software to visualise application-specific downloadable data (*URL 11.5*).

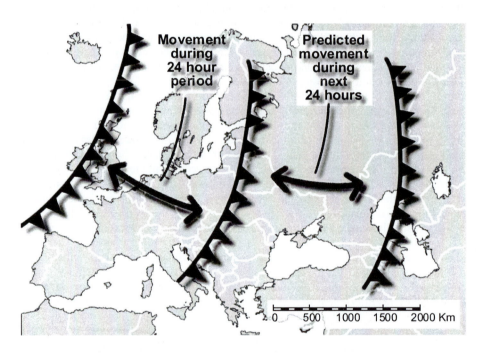

Figure 11.1 Space and time in a weather map.

With respect to the contents, the majority of the weather sites contains maps that show the condition of the atmosphere of the immediate past or present, based on observations at weather stations, on satellite imagery and/or on radar echoes. Also very common, of course, are forecasts of future conditions. Finally, there are sites offering access to archives of old maps, or to a variety of specific topics. They aim at atmospheric researchers, weather enthusiasts, pilots, farmers, people suffering from natural allergies, or others with specific interest in weather characteristics. Examples of the different types of weather maps are discussed in Section 11.3.

The weather map providers are almost as varied as their products, as will be made clear in this chapter. Many of the official national weather services disseminate maps on the Web. Other examples are satellite and radar data providers, some universities and schools. A number of weather forecasters or weather enthusiasts maintain sites, often with many links, e.g. Ed Aldus (*URL 11.6*) and Roger Brugge (*URL 11.7*). Among the commercial providers are newspapers and broadcasting organisations. Access to sites is sometimes restricted: membership, subscription or payment may be required (e.g. Flightbrief, *URL 11.8*). Some search engines have special weather pages, or "weather" as one of their predefined search categories, e.g. Yahoo! (*URL 11.9*), Excite (*URL 11.10*) and Hotbot (*URL 11.11*). Many links are also provided by Oddens' Bookmarks (*URL 11.12*) and by Yahooligans "for the kids" (*URL 11.13*).

After this introduction, the next section tries to characterise the Web as a medium for weather map dissemination and use. This is done by a comparison between the Web and conventional mass media, in particular television. The remaining part of the chapter deals with Web maps only. Four broad categories of weather maps are distinguished. In each category selected examples are described in more detail. The chapter ends with some expectations and hopes for the future.

11.2 THE MEDIUM: TELEVISION OR THE WEB?

Of the conventional mass media that provide weather information, the radio does not support graphical information, and the newspaper is a static medium. To illustrate what the role of the Web might be, comparison of television and the Web as media to provide weather maps seems to make sense. The main differences are summarised in Table 11.1.

Table 11.1 Summarised characteristics of weather map provision.

Characteristics	Television	Web
Narrative	always	not common
Explanation/interpretation	always	sometimes
Kind of information	general	general to highly specific
Access:		
• moment	on supply	on demand
• duration/frequency	time limitations	no time limitations
• interaction	none	low to high
Spatial coverage	limited	not limited
Up-to-dateness	reasonable	reasonable to very high

On television a real weather narrative is told. This is done by the presentation of items in a more or less fixed chronological sequence (immediate past, present, future), often supplemented by a stepwise focusing on smaller areas (e.g. from continental via national to regional levels). Complementary views add to the narrative (e.g. cloud cover images, maps showing the results of surface measurements, and 3D animations of moving air masses and associated fronts).

Furthermore, much explanation and interpretation are provided by the weather forecaster (Carter, 1996; Carter, 1998; Monmonier, 1999). Narratives are not common on the Web. Although the hyperlinked structure allows a user to "construct" a story, it usually requires some effort. Explanation of the map content varies, and interpretation of the weather phenomena that are represented is often completely lacking. Sometimes even essential elements such as legend and time indication are missing or illegible. On the other hand, a few sites offer abundant additional information: in text, graphics, animations, or by providing many links, as described below.

Only general weather information, of interest to a broad audience, is provided on television. Access to this information depends on supply, or broadcasting times, although there are also all-weather channels on cable in many countries. Other stations treat weather usually as a subset of the news, and the weather segment may be tailored to the time of day and the number of viewers (Carter, 1998). Also, the information is superficial: viewing times are brief, and replay on demand or other types of interaction are not offered. On the Web, weather information for different user groups, including highly specialised ones, is available. Access is on demand, there are no limitations in moment, duration and frequency of viewing other than the ones caused by traffic intensity on the Net, bandwidth, file sizes, etc. Maps can be downloaded or printed, and although there are "view only" weather maps, many maps offer some kind of interaction.

Other important characteristics of information provision are spatial coverage and up-to-dateness of the weather maps. On television, the maps are usually limited to the home country and continent of the broadcaster. During the holiday season more detailed information about popular foreign destinations might be provided. The up-to-dateness of the information is reasonable. Impressions of the weather "elsewhere" in the world can be easily obtained on the Web. The up-to-dateness of the information varies, from undefined to very high. (Near) real time views on the weather can be provided with continuous updates (e.g. by webcams; WetterOnline is one of the sites that provides such views for a number of German locations, like Hamburg, *URL 11.14*).

It can be concluded that television and the Web are complementary media, tailored to different types of use. Television offers easy "weather consumption": narrated, explained and entertaining. The weather segment is suitable for daily use if the area of interest is covered and general information is sufficient. The Web is more geared to provide answers to specific weather-related questions, it offers a broader range of weather visualisations, not tailored to some kind of "average" use. There are maps with familiar themes, but attention is also paid to medium and long range forecasts and to special phenomena, like visibility in the upper air (for pilots), ozone dynamics, severe weather, etc. There are tutorials and sites that allow clients to download data or visualisation software. In short, from a supply point of view the Web has to offer something to almost everyone, and a lot to someone with knowledge of, or interest in, weather phenomena. However, what kind of use people exactly make of weather maps, and whether the supply matches the demand has, as far as we know, not yet been investigated.

11.3 WEATHER MAPS ON THE WEB

In this section four categories of weather maps are distinguished. Maps in the first three categories, weather maps in a narrow sense, satellite image maps and radar image maps, describe recent or current conditions of the atmosphere, or they predict future conditions. Their sources (and appearances), however, are different. The fourth category, other weather and weather-related maps, contains maps of specific phenomena, including climatic maps. Each category is described separately by some general background information to provide the necessary setting, followed by a description of some examples. Since an objective, exhaustive search of the Web is difficult (if at all possible), selection is unavoidable (see also Harrower *et al.*, 1997). Selected are examples that are considered illustrative for the issues discussed in this book, if applicable to weather. It was found that North America is best-endowed with weather maps. Africa, for example, is much less favoured, and the maps that *are* found are often supplied by North American providers. Compare, for instance, the availability of maps for North America and Africa at CNN Weather (*URL 11.15*) and Yahoo! Weather (*URL 11.16*). Monmonier (1999) has already described a number of North American web map examples. The selections in this chapter are rather biased towards maps of Africa, although other illustrative examples are also provided.

11.3.1 Weather maps in a narrow sense

Weather maps showing the recent condition of the atmosphere are based on large amounts of data, measured at many meteorological observation points. In most cases ground measurements are made of elements like wind, temperature, humidity, precipitation, cloud cover, visibility and air pressure. Some posts also measure wind, temperature, humidity and pressure in the upper air. There is a worldwide network of meteorological observation posts. Individual countries have a central institute that collects the measurements in its area, and exchanges these in coded form with other countries. Exchange is essential, because weather is a global and very dynamic phenomenon. The national institutes use the Global Telecommunication System of the World Meteorological Organisation (WMO), the umbrella association, which also maintains a website (*URL 11.17*). Since it provides hyperlinks to its members with a website, this is a useful starting point for (virtual) travellers.

Measurements are taken at fixed times, although some observation posts sample more frequently than others. Plotting the point data measured at a particular time in a map enables overview for larger areas. The maps are analysed and integrated with other data (previous weather maps, radar and satellite imagery). Based on the analysis, meteorologists may add elements like isobars, fronts, high and low pressure cells and isotherms to complete the diagnosis for (a part of) a day. Point observations can also be aggregated to areas of mist, precipitation, etc.

For quite a number of countries, current weather maps are available through the central institutes (see the WMO site for links, *URL 11.17*). In most cases, these maps are static with temperature and pressure indications, icons, fronts, etc. In addition to the central institutes, there are a number of other providers. Intellicast (*URL 11.18*) offers a very limited choice for Africa, only temperature for the whole continent. The static map, although accompanied by a list of more locations, is clickable at very few locations only to view forecasts (not in maps). Yahoo! (*URL 11.16*) offers more choice: precipitation, high and low temperatures. The isoline maps are attractive: colourful and projected onto an image of part of the hemisphere (Figure 11.2 left). They are designed by Weathernews (WNI), a company that provides weather graphics to other disseminators (*URL 11.19*). USA Today (*URL 11.20*) also presents maps of WNI, but the design and part of the contents are different (Figure 11.2 right). Both examples are static view only maps. Weather Underground (*URL 11.21*) offers a relatively large choice of maps: temperature, visibility, wind, heat index, windchill, humidity and dewpoint. The bright and attractive static maps are interactive: details of current conditions and forecasts appear after clicking. Even more choice is offered at the extensive site of NOAA's Climate Prediction Center (CPC). CPC's African Products homepage (*URL 11.22*) provides access to various daily and seasonal wind, temperature and precipitation maps. Additional maps (except for North Africa) are encountered if the FEWS link is followed (*URL 11.23*). The maps are not designed for the Web (values are illegible), but acceptable as previews of downloadable data. Little explanation is provided, since the maps are meant for use by experts.

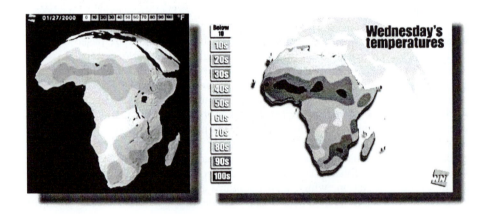

Figure 11.2 Current weather maps of Africa.
Left: high temperature, Yahoo! (*URL 11.16*); right: temperature, USA Today (*URL 11.20*).

On all these sites, many more weather maps are available for the USA, even at regional and local level. Most familiar are maps of the conditions at sea level, but there are also maps of higher levels in the atmosphere. The last group is particularly relevant for forecasting (e.g. of the direction and speed of movement of storms), because flows in the upper air can be strong, and they influence the conditions near the surface (see e.g. the upper air maps of UM Weather, *URL 11.24*). The patterns of flow, however, are usually simpler at higher levels, since there is not so much influence of terrain and temperature differences (Monmonier, 1999).

It can be safely assumed that most users of weather information are more interested in maps that deal with future weather than in maps that describe what has happened already. However, a good weather diagnosis is important for a prognosis. Forecasting starts with an analysis of present conditions and trends, followed by an approximation of future conditions and relations, using numerical weather prediction models. Many models are in use. Models for short and medium term forecasts (0–10 days) are of interest to users like travellers, or planners and executors of work that is somehow affected by the weather (farmers, painters, insurance companies, etc.). Meteorologists try to extend the outlook period further, because insight into the behaviour of weather elements over longer periods might improve their forecasting over shorter periods. The accuracy of forecast maps depends, of course, on the time for which the forecast is made. Other factors are the element of the weather depicted (e.g. precipitation is more difficult than temperature), the season and the area under consideration (Monmonier, 1999). Most models enable reasonable forecasts for three to five days. In addition to the short time span, the amount of detail is also limited.

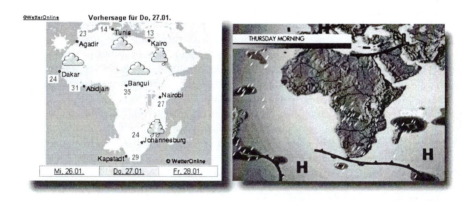

Figure 11.3 Forecast maps of Africa.
Left: Wetteronline (*URL 11.25*); right: CNN Weather (*URL 11.26*).

WetterOnline (*URL 11.25*) provides a simple forecast map of Africa (Figure 11.3)(left) and African regions, with temperature indication and icons (sun, clouds, precipitation) for a limited number of locations. A legend is not presented, and probably also is not necessary. At the bottom, links to forecast maps for two days further are provided. For a number of cities, some more details can be found by clicking in the maps. CNN (*URL 11.26*) provides a somewhat more sophisticated map (also without legend) for one day ahead (Figure 11.3)(right). On a base map containing relief and administrative boundaries, more or less transparent symbols for clouds and precipitation are shown. Also represented are high and low pressure centres, fronts and areas with strong winds. Temperature is not included. For temperature indications (four-day forecast in a table), one has to select a world region, then a world city from a list. The map itself is static and viewonly. CNN also has dynamic interactive maps for North America (*URL 11.15*), e.g. a forecast for the current day, for tomorrow and 24 hours temperature trends. The QuickTime movies are small, but clear with partly transparent weather symbols. Trajectories of fronts and showers can easily be followed. A thematic legend is not always provided, but the passing of time is indicated by a yellow transparent bar, that gradually shifts from left to right in viewing time. AccuWeather (*URL 11.27*) offers 3D fly-through animations of forecasts of the USA These animations probably look good on television, but are rather small and dark on the Web. They contain, for example, geographic names that are unfortunately barely legible. AccuWeather is also one of the providers that offers customised forecasts for various applications as one of their services.

Forecast maps for more than five days ahead are usually relatively vague due to the complexity of modelling. Models are complex because interactions in nature itself are complex and because many data are integrated in the modelling process, even data collected at a global scale, and for a number of levels in the atmosphere. These modelling results are usually presented at a really small scale. The content, especially of the long-range forecasts, is often limited to the probability that temperature and/or precipitation will be above, at, or below average values. ECMWF (*URL 11.28*) provides, for example, maps of Africa with precipitation, temperature and pressure anomalies for 1–3 months ahead.

Forecast models vary not only because of the time frame considered, but also due to the weather elements incorporated and their relative weighting, the assumptions made about future conditions and the influence of error propagation. UM Weather (*URL 11.24*) provides model selection from a number of possibilities for the USA It is also possible to select a period of validity and a specific layer of the atmosphere. The resulting forecast maps are not always clearly legible. Expert knowledge is required to interpret the maps. A prototype tool to assess the quality of the various models has been designed by researchers of the GeoVista Center of Penn State University, USA (Dirks Fauerbach *et al.*, 1996; *URL 11.29*).

There are also models that try to predict climatic changes, and models for special purposes, e.g. to predict storm tracks, or precipitation in the growing season. Examples of forecast maps resulting from these models are given in Section 11.3.4.

11.3.2 Satellite image maps

Satellite data are partly comparable with data collected on the ground. Hence, gaps in the network of meteorological observation points can, to a certain extent, be filled by satellite data, but even for areas with a dense network of points, this is a valuable resource. Also, data not measured on the ground can be acquired.

Energy reflected in the electromagnetic spectrum is used to detect a range of phenomena, which can be distinguished by characteristic spectral signatures. Best known, and available on the Web, are products generated from detection in the visible and infrared parts of the spectrum. On visible image maps, objects that are highly reflective appear bright and objects that have a low reflectivity appear dark.

So snow, ice and thick clouds are white, thinner clouds and mist are grey. At which level in the atmosphere the clouds exist cannot be determined from visible sources, the reflectivity is similar. Infrared image maps provide information on the temperature, and since the temperature decreases at higher levels in the atmosphere, high clouds appear brighter than low ones. In cloudfree areas, these image maps show the temperature of the Earth's surface. Desert areas, for example, are very dark if the data are collected during the day. The data are sometimes processed to colour infrared image maps to enhance temperature differences.

So distribution, height, thickness and temperature of the cloud top can all be determined. The Web is used to disseminate very up-to-date cloud cover image maps. From the description above, however, it can be concluded that it matters whether a user of image maps on the Web looks at visible or infrared products. A number of sites unfortunately fail to inform the user about this. Putting successive images in a loop provides additional information on direction and speed of movement. Analysis of spatio-temporal cloud patterns is very useful, because it enables recognition of phenomena such as depressions, cyclones, heavy showers and mist. If tornado-producing thunderstorms or other dangerous weather is detected, the American Goes satellites, for example, are able to zoom in and monitor the developments separately (Monmonier, 1999). The Goes satellites are part of an international network of satellites for worldwide, day and night cloud cover surveillance. Other information that can be derived from satellite data are water vapour and the distribution of atmospheric gasses and ozone (see also Section 11.3.4).

Figure 11.4 Java applet interface for dynamic maps (Weather Underground, *URL 11.30*).

For Africa, Yahoo! (*URL 11.16*) provides static (the most current map) and dynamic viewonly image maps from WNI (see Section 11.3.1). The clouds are colour enhanced (pink) but no base map information is added. Whether this is a visible or infrared image is unfortunately not indicated. CNN (*URL 11.15*) also provides static and dynamic viewonly image maps for Africa (from AccuWeather, see Section 11.3.1). A graticule and international boundaries are added for spatial reference. Again, no indication is given of the type of image. An interesting page is Weather Underground's tropical weather (*URL 11.30*), with satellite image maps for the Atlantic, Caribbean and the Pacific region. Various static view only visible and infrared cloud image maps are provided, but also dynamic infrared cloud and water vapour images maps. The Java applet interface used here (see Figure 11.4) offers more interaction than the QuickTime interface. It is, for example, possible to zoomin and to exclude images from the loop. Unfortunately it is difficult to read the temporal indications.

Satellite imagery is used for weather diagnosis, but also for the prognosis through forward projection of recent developments by computer processing. A special type of short term forecasting (less than 12 hours ahead, called "nowcasting") is sometimes applied by extrapolation of satellite observations. Meteo France (*URL 11.31*) provides current cloud cover images of France (see under "Observation/satellites"). Clicking on "Prévision/modèles" gives a forecast for 24 hours ahead (Figure 11.5). The static viewonly image maps are colour enhanced and international boundaries are added. Again, it is not indicated whether visible or infrared images are the source.

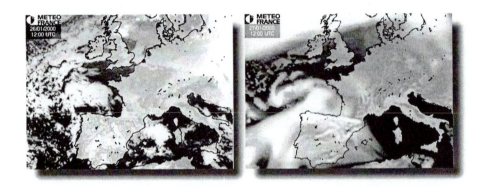

Figure 11.5 Satellite image maps of France and surroundings (Meteo France, *URL 11.31*).
Left: current status (observed at noon); right: predicted status for the next day (computer processed).

11.3.3 Radar image maps

Like satellite data, radar echoes can be used for weather diagnosis and prognosis, including nowcasting. But unlike the sensors of a satellite, a radar sensor actively radiates energy (from a ground station). Energy that encounters objects is reflected in all directions. Part of it, the backscatter, is again detected by a receiver. Different types of storms, precipitation, clouds and dust can be recognised from the echoes because they reflect in different parts of the microwave band, and their reflections differ in shape and brightness. Putting the image maps sequentially in a loop, clearly shows the path and intensity of blowing showers. Usually the number of radar image maps in a loop is greater than the number of satellite image maps in an animation. Doppler radar is particularly suitable for short-term storm forecasts and warnings, because it enables quick differentiation between storms that are moving towards the radar (short and high frequency waves) and storms moving away from it (longer, low frequency waves). For this reason, Doppler radar stations are located at strategic positions in parts of the United States that are frequently hit by tornadoes (Monmonier, 1999). Radar is useful, but very expensive. It cannot effect-
tively be used in rough terrain, and even in flat terrain the spatial range is limited.

Radar image maps for Africa are hard to find. For the United States, however, these image maps are available at national, regional and even at local level, in static and dynamic form (e.g. The Weatherlabs, *URL 11.32*). An interesting page can be found at NCDC's site *(URL 11.33)* because it allows much interaction with the data. Selections can be made for the duration of the period (day, month, year), the interval (1 hour, 3 hours, 1 day) and the zoom factor (2–5 times). The system then represents twelve radar mosaics in a loop. On a black background with white administrative boundaries, the radar reflections are represented in spectral colours. A thematic legend is provided, temporal indications are in text, which is difficult to follow together with the moving images. A temporal legend, (e.g. like CNN uses for forecast animations, see Section 11.3.1) would be an improvement here. A European example of Doppler radar is the route of a severe hail shower that hit Lucerne, Switzerland on July 21st, 1998 (Figure 11.6). The base map shows a very nice depiction of the relief and other topographic elements, over which the transparent shower moves. The lack of geographic names, however, makes it difficult to locate the shower in space (ETH, *URL 11.34*).

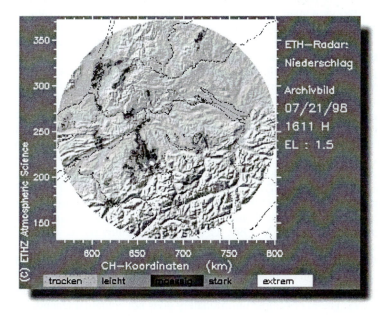

Figure 11.6 Doppler radar animation: severe hail over Lucerne, Switzerland, 21.07.1998
(ETH, *URL 11.34*).

The images available on the Web are usually compositions of echoes from a number of stations. Even for a small country like the Netherlands, the precipitation radar image maps, disseminated every 15 minutes, are composed from measurements at two stations *(URL 11.35)*.

11.3.4 Other weather and weather-related maps

There are many maps of atmospheric phenomena that do not fit well into the categories discussed above. Examples include displays from archives containing weather data or maps that are older than the immediate past, maps showing a synthesis of weather conditions that have prevailed over an area during a long time (climatic maps), and maps of specific phenomena, like pollen forecasts, severe weather or the monsoon. The examples described below are just a selection from this last broad category.

Archives that can be accessed are provided by Weather Underground (*URL 11.21*) for Africa. The clickable maps described in Section 11.3.1 above enable the user to request historical conditions up to a year back from the current date for cities in a number of countries. CPC is another example: its stratosphere homepage provides the latest ozone image maps, and access to archives, even of animated GIF files (*URL 11.36*).

CPC also provides *climate* outlooks for the USA (*URL 11.37*) Current monthly/seasonal forecasts is a page with a number of small scale maps, accompanied by a legend with probabilities of occurrence for each class. The user has to scroll to view map and legend. CPC offers much more, for example threats caused by temperature, precipitation and wind, including drought and wildfire dangers. These are derived from long-term forecasts. Worth mentioning is further NCDC's Climatic Visualization System (CLIMVIS) (*URL 11.3*). It enables visual browsing of the online available data for the USA It offers a choice from a number of parameters, over long periods, of which plots can be made.

Examples of specific phenomena can be found at Yahoo! (*URL 11.16*), where pollen forecasts for the USA are available all year round. The source of the pollen and the potential severity of the allergy problem are represented at national, regional and local level. More seasonal examples on other sites are UV and snow forecasts, or "Fall foliage", showing the time of peak autumn colours for regions in the USA (The WeatherLabs, *URL 11.38*). Other specific phenomena are hurricanes and storms. NCDC (*URL 11.39*) offers a large number of static and a few dynamic image maps (MPEGs). A nice example is the false colour sunset image (with administrative boundaries) of hurricane Floyd, approaching the Bahamas on September 12, 1999. CNN's storm centre (*URL 11.40*) offers a VRML-view of a hurricane, with an animation running in it. Interpretation is not easy. NASA's Data Assimilation Office (*URL 11.41*) offers 2D and 3D dynamic maps of the Asian monsoon. In the 2D example, average wind (April 1985–December 1989) is represented by vectors, and precipitation by a hue-value scale (Figure 11.7). The 3D example unfortunately does not contain a legend, but shows relief of the coasts, the ocean surface with (most likely) precipitation levels, wind vectors, and ribbons representing trajectories of air parcels. Many things are happening in these animations, particularly in the 3D example, and although the QuickTime format allows limited interaction, more interaction and explanation would certainly be helpful here. On the same site (under atmospheric chemistry/images) QuickTime movies of the Antarctic ozone hole are provided. In general, it was found that many of the QuickTime examples on the Web are too small, dark and vague to be legible. Two sites that are interesting from an educational point of view are CPC (*URL 11.42*) for its many maps and the tutorial on the El Niño/La Niña phenomenon, and the University of Cambridge (*URL*

11.43) for its animated ozone tour. Animations are provided in several formats. A final example is provided by an animation of the eclipse of August 11 1999 over Europe. The view only animation contains images for every 10 minutes between 08.10 and 12.40 hrs, on which the shaded area is gradually moving to the Southeast (Eumetsat, *URL 11.44*).

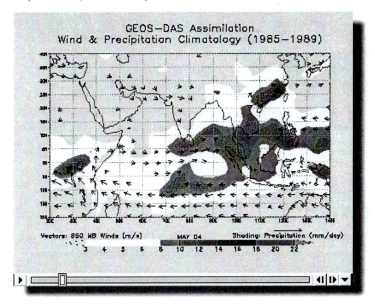

Figure 11.7 QuickTime animation of monsoon parameters
(NASA's Data Assimilation Office, *URL 11.41*).

11.4 EXPECTATIONS AND HOPES

Dissemination of weather maps on the Web will, no doubt, change. Next to expectations, there are some hopes for the future. Both will be briefly discussed.

Expectations can partly be derived from trends that are already visible. One expectation is a rapid improvement in the design of weather maps on the Web. Over a few months of searching, it became clear that the number of scanned paper and other maps not specifically designed for the Web is rapidly decreasing. Frequent change and up-to-dateness, so important for weather information dissemination, also applies to the content of the weather sites. Other expectations can be broadly assigned to technological, professional and commercial developments. Among the technological developments are diminishing download speeds, which is important for weather animations. Also, more intelligent browsers might be able to select and represent the kind of information in which a specific user is interested (Johnson and Gluck, 1997). Furthermore, information is becoming available at higher spatial resolutions, which will result in more detailed weather information, even at a local level. Expected professional developments are more integration of different kinds of data and more interaction with the data and the display. The National Weather Service of the USA integrates hydrological and meteorological data from different sources *(URL 11.45)*, while GODDARD DAAC provides interdisciplinary data sets *(URL 11.1)*. A good example of interaction with the data, where the client is able to compose a representation, is already provided by NCDC (for CLIMVIS and radar data, *URL 11.3*). A type of interaction where various scientists at different locations are simultaneously working on the visualisation of data (collaborative visualisation) might be very suitable for complex weather and weather-related data. Finally, commercialisation will grow. Customised weather information can already be uploaded to the home page of clients of weather information providers (e.g. Weather Underground's weather stickers, see Figure 11.8, *URL 11.46*; The WeatherLabs, *URL 11.47*). These providers can also deliver the required information by e-mail, which will be more and more channelled through wireless devices via the Wireless Application Protocol (WAP), e.g. WetterOnline *(URL 11.48)*. Maps are not commonly included yet, but they might well be in the immediate future.

Figure 11.8 Customised weather sticker for Airbase Twenthe, the Netherlands
(URL 11.46).

An important hope is that more attention will be paid to the accessibility of the contents of the maps. On the one hand, it simply means providing full explanation of the image source and of the thematic and temporal contents (by clickable symbols or legends, including temporal ones). On the other hand, more explanation and interpretation of the weather phenomena itself would make many displays more accessible to interested laymen. Good examples are already existing: USA Today's *Hows and whys*, with explanation of weather phenomena by mouse-over graphics accompanied with text *(URL 11.49)* and *Ask Jack*, with a weather FAQ *(URL 11.50)*. Another example is the tutorial about the El Niño/La Niña phenomenon (the ENSO cycle), with many maps, diagrams and links offered by CPC *(URL 11.42)*. Other examples are the online guides (e.g. to meteorology and to reading weather maps) of the University of Illinois *(URL 11.51)* The hyperlinked structure of the Web enables (literal) presentation of information at different levels of depth to accommodate different knowledge levels of the users.

URLs

URL 11.1   Goddard DAAC data sets
                      <http://daac.gsfc.nasa.gov/data/dataset/index.html>
URL 11.2   NCDC <http://www.ncdc.noaa.gov/>

URL 11.3 NCDC CLIMVIS
<http://www.ncdc.noaa.gov/onlineprod/drought/xmgr.html>

URL 11.4 UM Weather software archive
<http://cirrus.sprl.umich.edu/wxnet/software.html>

URL 11.5 Meteo Consult Weerbeeld
<http://www.meteocon.nl/diensten/consumenten/weerbeeld/index_
weerbeeld.html>

URL 11.6 Ed Aldus European weather site <http://www.weather.nl>

URL 11.7 Roger Brugge <http://www.met.rdg.ac.uk:80/~brugge/>

URL 11.8 Flightbrief <http://www.flightbrief.com/>

URL 11.9 Yahoo! Weather <http://weather.yahoo.com/>

URL 11.10 Excite Weather <http://www.excite.com/weather/>

URL 11.11 Hotbot Weather <http://dir.hotbot.lycos.com/News/Weather/>

URL 11.12 Oddens' Bookmarks <http://oddens.geog.uu.nl/index.html>

URL 11.13 Yahooligans Weather
<http://www.yahooligans.com/Science_and_Nature/The_Earth/Weather/>

URL 11.14 WetterOnline Aktuelles Wetter
<http://www.wetteronline.de/de/aktuell.htm>

URL 11.15 CNN Weather maps and images
<http://cnn.com/WEATHER/images.html>

URL 11.16 Yahoo Weather images <http://weather.yahoo.com/images.html>

URL 11.17 WMO <http://www.wmo.ch/web-en/member_g.html>

URL 11.18 Intellicast <http://www.intellicast.com/LocalWeather/World/Africa/>

URL 11.19 WNI <http://us.weathernews.com/sales/>

URL 11.20 USA Today Weather
<http://www.usatoday.com/weather/basemaps/waft1.htm>

URL 11.21 Weather Underground
<http://www.wunderground.com/global/AF_ST_Index.html>

URL 11.22 CPC African desk
<http://www.cpc.ncep.noaa.gov/products/african-desk/index.html>

URL 11.23 CPC Fews <http://www.cpc.ncep.noaa.gov/products/fews/index.html>

URL 11.24 UM Weather computer model forecasts
<http://cirrus.sprl.umich.edu/wxnet/model/model.html>

URL 11.25 WetterOnline Forecasts Africa
<http://www.wetteronline.de/de/afrika.htm>

URL 11.26 CNN Africa forecast map
<http://cnn.com/WEATHER/Africa/forecast_map.html>

URL 11.27 AccuWeather products and services
<http://www.accuweather.com/wx/services/flythrus.htm>

URL 11.28 ECMWF seasonal forecasting
<http://www.ecmwf.int/services/seasonal/forecast/index.html>

URL 11.29 GeoVista Center: visualisation of uncertainty
<http://www.geovista.psu.edu/ica/icavis/febm/sdhbivar.html>

URL 11.30 Weather Underground tropical weather
<http://server2.as5000.com/tropical/>

URL 11.31 Meteo France
<http://www.meteo.fr/temps/france/satellite/sous_panneaux.html>

URL 11.32 The Weatherlabs Doppler radar <http://www.weatherlabs.com/cgi-
java/WL_WLServer?templatename=templates/usa/doppler.html&mode=E>

URL 11.33   NCDC Nexrad <http://www4.ncdc.noaa.gov/cgi-win/wwcgi.dll?WWNEXRAD~Images2>

URL 11.34   ETH Hagel <http://www.radar.ethz.ch/archiv_980721.html>

URL 11.35   De Weerkamer radar <http://weerkamer.nl/radar/>

URL 11.36   CPC stratosphere <http://www.cpc.ncep.noaa.gov/products/stratosphere>

URL 11.37   CPC forecasts <http://www.cpc.ncep.noaa.gov/products/forecasts/>

URL 11.38   The Weatherlabs Fall foliage <http://www.weatherlabs.com/cgi-java/WL_WLServer?templatename=templates/usregion/springoutlook.html&mode=E>

URL 11.39   NCDC hurricanes <http://www.ncdc.noaa.gov/ol/satellite/olimages.html>

URL 11.40   CNN hurricanes <http://cnn.com/SPECIALS/multimedia/vrml/hurricane/>

URL 11.41   Data Assimilation Office <http://daac.gsfc.nasa.gov/CAMPAIGN_DOCS/atmospheric_dynamics/ad_images_dao_animations.html>

URL 11.42   Climate Prediction Center Enso cycle <http://www.cpc.ncep.noaa.gov/products/analysis_monitoring/ensocycle/>

URL 11.43   University of Cambridge <http://www.atm.ch.cam.ac.uk/tour/index.html>

URL 11.44   Eumetsat <http://www.eumetsat.de/en/index.html?area=left3.html&body=/en/area3/metenews/eclipse_eur.html&a=316&b=2&c=310&d=300&e=0>

URL 11.45   National Weather Service AWIPS <http://205.156.54.206/msm/awips/awipsmsm.htm>

URL 11.46   Weather Underground weather stickers <http://www.wunderground.com/geo/BannerPromo/global/stations/06290.html>

URL 11.47   The WeatherLabs: Weather on your site < http://www.weatherlabs.com/cgi_java/WL_WLServer?templatename=templates/nogeoloc/weatheronyoursite.html&mode=E>

URL 11.48   WetterOnline Wap-Angebot <http://www.wetteronline.de/cgi-bin/fcgi/step.fcgi?URL=/feature/de/wap.shtml&STEP=wpwetter>

URL 11.49   USA Today Hows and whys <http://www.usatoday.com/weather/wgraph0.htm>

URL11.50   USA Today Ask Jack <http://www.usatoday.com/weather/askjack/wjack1.htm> enso_cycle.html>

URL 11.51   University of Illinois Online guides <http://ww2010.atmos.uiuc.edu/(Gh)/guides/mtr/home.rxml>

REFERENCES

Carter, J.R., 1996, Television weather broadcasts: animated cartography aplenty. In
Proceedings of the seminar on teaching animated cartography, Madrid, edited by Ormeling, F.J., Köbben, B. and Perez Gomez, R., (Utrecht: International Cartographic Association), pp. 41-44.

Carter, J.R., 1998, Uses, users, and use environments of television maps. *Cartographic Perspectives,* (30), pp. 18-37.

Dirks Fauerbach, E., Edsall, R.M., Barnes, D. and MacEachren, A.M., 1996, Visualization of uncertainty in meteorological forecast models. *Geovista Center*

<http://www.geovista.psu.edu/ica/icavis/febm.html> (accessed June 28, 2000).

Harrower, M., Keller, C.P. and Hocking, D., 1997, Cartography on the Internet: thoughts and a preliminary user survey. *Cartographic Perspectives*, (27), pp. 27-37.

Johnson, D. and Gluck, M., 1997, Geographic information retrieval and the World Wide Web: a match made in electronic space. *Cartographic Perspectives*, (26), pp. 13-26.

Monmonier, M., 1997, The weather map: exploiting electronic telecommunications to forecast the geography of the atmosphere. In *Ten geographic ideas that changed the world*, edited by Hanson, S., (New Brunswick: NJ, Rutgers University Press), pp. 40-59.

Monmonier, M., 1999, *Air Apparent, how meteorologists learned to map, predict, and dramatize weather*, (Chicago: University of Chicago Press).

CHAPTER TWELVE

Web maps and road traffic

Nicoline N.M. Emmer

12.1 INTRODUCTION

Traffic is a very broad concept. It encompasses many types of traffic, e.g. railway and telephone traffic. This chapter will, however, focus on one particular type of traffic, namely travelling by car or other motorised vehicles.

The following section introduces the topic and indicates the advantage to motorists of using the WWW, particularly if maps are provided. This is followed by a description of the types of information given on traffic sites and how this information is presented and graphically visualised to the end user (Section 12.3). The general needs of motorists are also investigated here. Based on the these needs, possible improvements to current traffic sites and especially to their graphics, are suggested (Section 12.4). The final section gives a brief overview of future developments and cartographic innovations that can be useful for traffic web maps.

12.2 WHY USE TRAFFIC WEB MAPS?

When travelling by car interesting questions are: "I want to drive to location x, how do I get there? What is the shortest or quickest route? How long will it take me to get there?" Travelling by car is all about connections between point locations, distance and the time taken. Many information sources provide information on these questions, such as radio, teletext, telephone, television, CD-ROM route planners, e-mail (*URL 12.1*), etc. So why should the Internet be used to provide the answers to traffic questions? Or postulated differently: "What profits does the Internet yield in comparison to other information sources?"

Road traffic is extremely dynamic: information on traffic jams or closures changes every hour. Information about this kind of data could be mapped with "conventional" production methods, but the dissemination of data was a problem. Before the information was distributed, the data were already outdated. With the arrival of the Internet this problem is solved: it is possible to display real-time data. The Internet offers more advantages. Firstly, it is very accessible. For example, an end user can obtain information about traffic conditions at any time anywhere as long as he possesses a computer with access to the Internet. The Internet is platform independent, so communication is possible between different operating systems. In addition, the Internet can supply interactivity, allowing the end user to be in direct contact with the data. This interactivity allows the user to decide what type of information is to be provided. Travellers can therefore customise the information to fit their own needs. An additional advantage of the Internet is that much of the information is provided free of charge to the user (often in fact paid for by advertisements).

To sum up, the Internet is suited to provide real-time information on traffic, including in map form. Because of the Internet's functionalities of interactivity, display of real-time data and accessibility this chapter will focus on website traffic maps that make use of these functions. Thus maps that offer no additional information compared to paper maps are not dealt with.

12.3 CURRENT DISSEMINATION OF TRAFFIC INFORMATION

12.3.1 Nature of road traffic sites

Sites providing information about travelling by car can roughly be divided into two categories: sites offering routing facilities, and sites providing information on traffic conditions.

Most sites providing routing services are commercially oriented. For example mapping engines like MapQuest (*URL 12.2*), MapBlast (*URL 12.3*) or DeLorme CyberMaps (*URL 12.4*) offer routing facilities. The companies owning these mapping engines offer these tools to others which imbed the tools in applications like search engines. For example the search engines Yahoo!, HotBot, Lycos, and WebCrawler offer routing information by using the TripQuest utility from MapQuest. Also corporations like Shell (*URL 12.5*) and Michelin (*URL 12.6*) provide facilities to plan a route. The maps offered by routing applications originate from mapping corporations like AND (*URL 12.7*), Navigation Technologies (*URL 12.8*) (McCutcheon, 1999), Vicinity Inc. (*URL 12.9*), or Geographic Data Technology Inc. (*URL 12.10*). In general, routing facilities are supplied free of charge, although in some cases a membership is required, e.g. RouteMaster by the AAA (*URL 12.11*). Routing utilities provide end-users with information on route description, fuel consumption, shortest route, quickest route or estimation of total distance and travel time. Most of these aspects are not time sensitive, except for travel time estimates that depend on current traffic conditions. Routing facilities do not (yet) base their travel time calculations on current traffic conditions, although some sites do provide direct links to current traffic information, e.g. MapQuest.

Sites providing data on traffic conditions in general deal with two groups of information: incidents and traffic density. An indication of traffic density is given by providing information on e.g. traffic speed, traffic congestion, traffic volume, speed traps, etc. Construction, closures, accidents, bad weather conditions, etc. in general are grouped under the heading incidents. Traffic density and incidents are both time-sensitive aspects. Terms associated with these types of sites indicate this: real-time traffic information, traffic view, traffic watch, etc. Websites offering time-sensitive data mostly have an experimental character and are owned by governmental institutions like Departments of Transportation and Automobile Associations like the American AAA and the Dutch ANWB (*URL 12.12*). Very often information on public transport is also provided by these kinds of sites.

12.3.2 Visualisation of routing facilities

Within routing applications a route is defined by submitting start and destination points. It may also be possible to insert several intermediate points or to avoid areas or particular roads, e.g. toll roads, or to define the type of route, e.g. shortest, quickest or most scenic. Marking out a route, by clicking on the map, is not supported by route planners. Some applications can pick a route on city name, street name or even on the street address. Along with filling in start and destination points the user can decide upon the type of output. Most often one overview map is given, but sometimes when a long distance route is required more maps are available. In some cases two detailed maps of the start and destination points are incorporated. Beside the overview map the user can in general choose additional information in the form of text or turn-by-turn maps (Figure 12.1). Turn-by-turn maps have a larger scale than the overview map and can be considered as static view only maps.

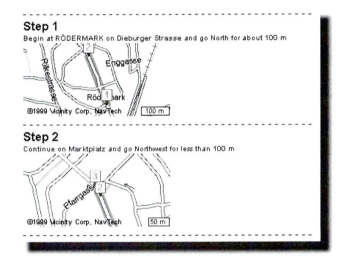

Figure 12.1 Turn–by–turn maps (*URL 12.5*).

The content of routing maps is, just like paper maps, determined by scale. It is noteworthy that within many routing engines maps of different scales are provided by different mapping organisations. In general the map image can be considered primitive. Interactive functionality of maps is offered through pan and stepped zoom options; in general maps are not clickable. The degree of interactivity depends on the organisation that provides the routing facility. In general the maps can be customised to some degree. For example map size, map units, map colours, or icons can be altered. Specific functionalities, such as measuring tools, offered by CD-ROM versions of route planners are not incorporated in routing sites. On the other hand, routing sites offer other utilities. For example, the mapping machine "CyberRouter" of DeLorme provides a facility

one can go within 15 minutes. This is based on current traffic speed. A nicely designed map is the DRIVES demo page of Globis Data Inc. (*URL 12.24*).

Many good examples of "dynamic interactive" traffic maps exist. In the United Kingdom there is Vauxhall Traffic Net (*URL 12.25*), while in the Netherlands the motorist and cyclist organisation ANWB (*URL 12.26*) provides interactive real-time maps. For the USA examples include Current San Jose Traffic (*URL 12.17*), and the Gary-Chicago-Milwaukee (CGM) Corridor Home Page (*URL 12.27*). At the San Jose site the contents of the map may be altered. The user can choose information on speed, volume and congestion. The Icelandic Vegagerdin site (*URL 12.28*) offers another way to inform users about traffic density. In a permanent frame on the map the number of vehicles since midnight are displayed for certain roads. At the Vauxhall Traffic Net site a dynamic way of representing information on traffic speed is applied. A blinking coloured arrow (point symbol) shows the location where speed is below 25 mph. The colour of the arrow symbolises a certain speed. The same principle is applied at the CGM Corridor Homepage and the American Tennessee Department of Transportation site (*URL 12.29*). The latter shows weekly information on interstate construction. The map itself can be regarded as "dynamic interactive". Although the map does not show real-time data the map is considered to be dynamic because the data are presented through "blinking". Further these incident locations are numbered by orange squares. The numbering refers to text underneath giving more details on that particular construction project. These orange squares are also clickable. This hotspot brings the user to the relevant part of the text.

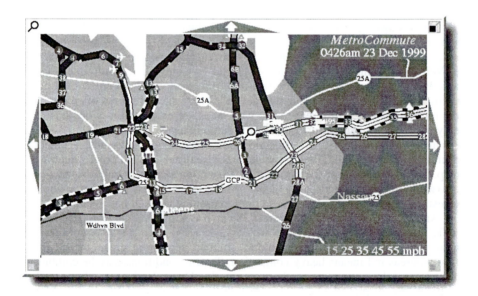

Figure 12.3 MetroCommute site (©MetroCommute.com, LLC2000) (*URL 12.30*).

The above mentioned sites have limited or no zoom or pan functions. Examples, however, do exist where advanced zoom and pan functionalities are applied. For example the road avoidance guide by Startribune, in the Minneapolis – St. Paul region *(URL 12.31)*, uses a Flash plugin to zoom and pan. Even though the technique is applied in a simple form, it shows the current technical capabilities. A more sophisticated example of zoom and pan functions is offered by the New York regional traffic information site called "MetroCommute" (see Figure 12.3). Through the upper two corners of the map it is possible to increase or decrease the scale of the map. In the lower two corners functionality to enlarge or decrease the size of the map is available. Panning options are offered through arrows on the sides of the map. This map offers interactivity that is not very often applied, namely interactivity through mouse-over events. When a hotspot is activated by a mouse-over a small text display with current traffic speeds appears. This map also contains a time stamp that indicates how up-to-date the map is.

Not only is traffic information visualised through maps. Information on traffic conditions is also gathered via video cameras. Here, maps also play a dominant role. Locations of video cameras are displayed on maps by means of icons. These icons show, when activated by a click of the mouse, a video image of that location. The cameras provide a bird's eye view of traffic along major freeways or roads. The video page will automatically refresh the screen every minute or so. This means that the application can be regarded as real time. In some cases live video is transmitted to the Internet user. This, however, calls for a special plugin like RealPlayer *(URL 12.32)*. The Urban Traffic Control Centre *(URL 12.33)* shows television images of traffic in Belfast. Because it is not always very obvious in which direction the cameras are pointing reference images are used for orientation. Video images always contain a label containing time and location. Although these images are applied very often, one can wonder if ordinary motorists really need them.

12.4 IMPROVEMENTS OF TRAFFIC SITES

12.4.1 The user perspective

Because maps are about communication it is important that map design is tailored to user needs. Therefore, it is important to know what users require and how, from this perspective, sites, and especially maps, can be improved. When taking into consideration user needs four interesting questions emerge:
1.   What kind of information is desired?
2.   Where does the user want to receive the information?
3.   When does the user want to receive the information?
4.   How should the information be presented?

The desired information depends on the characteristics of the user. Different types of user can be differentiated on the basis of travelling motive. In the Netherlands half of the motorists on the highway are commuters, one third of the users travel for business reasons and the rest travel for social or recreational reasons. These different types of users require different types of information (Ministerie van Verkeer en Waterstaat, 1996). It appears that motorists are interested in information on traffic conditions only, or they want to be advised about the optimal route based on current traffic conditions. Furthermore, the users require that this information, on the above aspects, be accurate, reliable and up to date. Additionally, the need for information increases as the importance of the journey increases, especially in the case where alternative routes are available during periods of heavy congestion (Orski, 1997). Commuters are only interested in traffic conditions, especially the location, length and cause of traffic jams. Because commuters drive the same route every day and they are aware of possible alternatives, only information on traffic dynamics is required. They then choose from the alternatives based on the information provided and on their own estimate of how traffic conditions will develop. Business, social and recreational motorists are in general interested in optimal routes, travelling time, arrival time, and delay time because of unfamiliarity with that particular route. This type of motorist wants to be advised about route alternatives.

Concerning the place where a traveller wants to receive the information it is clear that always up-to-date information is required. This means that the information should primarily be provided while actually driving. However, it appears that users also desire information at other times (Ministerie van Verkeer en Waterstaat, 1996). There is a need for acquiring data at home, at the office, and at stopping points during the journey. Examples of the last situation are retrieving information while filling up the fuel tank or while waiting at a traffic light. To conclude, users want to receive information anytime and anywhere.

The purpose of providing information on traffic conditions and routing possibilities is to support the end user in deciding which route is the best, given a particular time. The presentation method should support this. The information can be supplied in text, audio, or graphical form. The presentation method is determined by the type of user (user needs) and using circumstances (time and location). These aspects can differ considerably and therefore the presentation methods should differ too. Luckily, many ways exist to present information to users. The Internet is one way, and offers many possibilities.

The Internet as a source for traffic information is not yet widespread. A research on commuter response to real-time traveller information found out that 14% of commuters are aware of the fact that they could obtain current traffic information through their computers or over the telephone (Orski, 1997). Hence only some make use of the services available on the Internet. It seems reasonable to conclude that if traffic websites are to play a greater role in providing information they should support those techniques for which the Internet is unique: real-time graphics and interactivity (see Section 12.2). Fortunately, due to rapid developments in telecommunication technologies the Internet can play a prominent role in providing these services to different users under different reading circumstances. The arrival of a Wireless Application Protocol (WAP) is promising in this respect (see Appendix A). The WAP is a specification for a set of communication protocols that standardises the way in which wireless devices, such as cellular telephones and radio, can be used for Internet access, including e-mail (Whatis.com, 2000). This option really opens up new markets. It was already possible to download driving directions to a PDA, but now it is also possible to have direct access to the Internet. This means that motorists can have access to the Internet anywhere and at any time.

The time needed to acquire the information should be of short duration. After all, the user checks for information before leaving home or work and during the journey. This means that the information must be available immediately. The end user should not be forced to click too much to obtain the final information. Therefore, all the information should be available on one page, and should be obtained in one overview. This time constraint is, however, not so important for those travelling for social or recreation purposes.

12.4.2 The design perspective

If the answers to the questions postulated in Section 12.4.1 are taken into consideration a few requirements emerge. It is desirable that routing facilities integrate information on traffic conditions within the service. Based on current and future conditions, routing applications should calculate the optimal route. Because nowadays it is possible to have access to web maps at any time and anywhere, sites must offer flexible functionalities according to reading circumstances. While driving a car it is preferable that information on route and traffic conditions is transferred via audio media. In other cases, maps and textual information are preferred. This means that the interface and map image should be transparent and not too complex. This in turn requires careful design of the web map (see Chapter 7), in order to ensure an efficient and rapid transfer of information.

The map structure must be well designed with full use of interactivity. For example, being able to switch layers on and off enables easy access to different aspects. Although this is already possible at several sites, this functionality can be applied more efficiently. While designing a map one has to take into consideration the effect on the map image when more layers are switched on at the same time. Next, the map should be displayed in its entirety on screen so that the user should not have to scroll. This also implies that easy zoom and pan functions should be available. The way additional information is given is not very efficient in some sites. Very often a new page or navigator window is opened to display the new information. It is much better if the information is displayed on the original page, preferably by mouse-over or in a different frame. More usage should be made of mouse-over events because the information is direct and only available when wanted. Mouse-overs can also take over the functionality of all or part of a legend (see Chapter 5). If legends are applied, their contents should be unambiguous, consistent and well structured, with a clear hierarchy.

Symbol design on traffic sites, especially on incident aspects, is in general very poor, with little consideration given to the proper use of the graphic variables (see Chapter 5). For example, aspects of traffic conditions most often are of an ordinal nature, preferably displayed by using a value scale. Nevertheless, very frequently different colours (hues) are used, because of their strongly associative characteristics. But this approach can very soon lead to confusion if different ordinal traffic conditions are all displayed in one map, with many different hues in total. Best is to use the variable hue only to distinguish among the different categories. Other design deficiencies include symbols that are too numerous and too large compared to the whole map (*URL 12.34*), leading to the map image becoming cluttered. In the case of pictorial symbols, if they are clear and easy to read on the screen they may be too large compared to the map while if they are reduced in size they may no longer be clear. Great care is needed to design reasonably small yet clear pictorial symbols (see Section 7.3). It is common to find on web maps that the concept of visual hierarchy is ignored with the background being as dominant as the subject itself. Very bright colours may be used and it may even occur that less important features are represented by brighter colours than the important features. A similar problem may exist with regard to symbol sizes and text on the map.

Many sites do not take into account navigational features, such as important buildings, built-up area and so on. In case of the ANWB traffic information site and the Athens real-time traffic map (*URL 12.23*) landmark buildings are displayed as oblique views (see Figure 12.4). The addition of navigational features can support a quick orientation of the driver and this adds to the readability of a traffic web map.

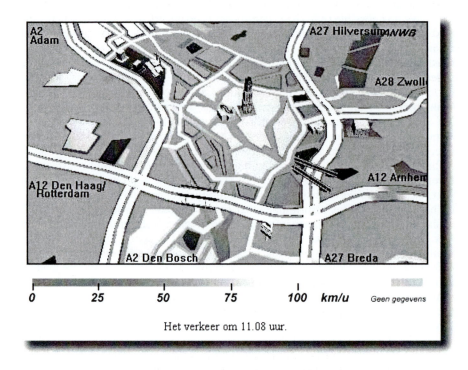

Figure 12.4 Application of oblique views of buildings on the ANWB traffic information site (*URL 12.26*).

12.5 FUTURE DEVELOPMENTS

Two developments are likely to influence traffic web maps in the near future: WAP and VRML. The ability to have direct access to traffic sites on mobile devices will determine the look of web maps. These mobile devices have very small displays, and this puts constraints on the map and on the user interface. The content of the maps and its design must be very simple. Only essential information should be displayed. Furthermore, the displays of PDAs are not so advanced as computer screens. For example some PDAs only have a black and white screen which puts some limitations on the map design

A virtual drive-through can contribute to presentation of information on traffic conditions. A motorist should have the possibility to enter his start and destination point before leaving home. Subsequently the computer calculates the quickest or shortest route based on current traffic conditions. A VRML model is produced of this route, which the user can download. Through this model different traffic conditions can be simulated: congestion, fog, storm and so on. This makes navigation during the actual drive easier, because the end user is now acquainted with the environment and obstacles. A driver should be able to manipulate the path of a drive-through so that alternative routes can be examined. On the "GoHamptonRoads" site it is already possible to virtually explore a predefined route (*URL 12.35*). Several video images of critical stretches of the chosen route are displayed. Of course, one always should ask oneself if users really desire a virtual tour of the route. In fact, there still are many unanswered questions concerning the real needs of the different users of road traffic maps, with definite scope for further user research.

URLs

URL 12.1   Twin Cities Express Traveler Information
               <http://twincitiesexpress.com/travelerinformation.asp>
URL 12.2   MapQuest <http://www.mapquest.com/>
URL 12.3   MapBlast <http://www.mapblast.com/mblast/index.mb>
URL 12.4   DeLorme Cybermaps <http://www.delorme.com/cybermaps/>
URL 12.5   Shell GeoStar <http://www.shellgeostar.com/home/default.asp>
URL 12.6   Michelin – Your route
               <http://www.michelin-travel.com/eng/itin/demande/amieponctuel.cgi>
URL 12.7   AND Internatinal Publisher. <http://www.and.com>
URL 12.8   Navigation Technologies <http://www.navtech.com/index.html>
URL 12.9   Vicinity Corporation <http://www.vicinity.com/>
URL 12.10 GDT Geographic Data Technology
               <http://www.geographic.com/index.cfm>
URL 12.11 American Automobile Association
               <http://www.aaa-calif.com/auto-travel/directions/index.asp>
URL 12.12 ANWB City <http://www.anwb.nl/city/anwb.htm>
URL 12.13 Avantgo <http://www.avantgo.com>
URL 12.14 Falk routing <http://www.falk-online.de/go_routing.html>
URL 12.15 Puget Sound Traffic Conditions
               <http://www.wsdot.wa.gov/PugetSoundTraffic/>
URL 12.16 Chicagoland Expressway Congestion Map
               <http://www.ai.eecs.uic.edu/GCM/chicagoland-java.html>
URL 12.17 Current San Jose Traffic
               <http://www.ci.san-jose.ca.us/traffic/sj_downtown_512.html>
URL 12.18 Gary–Chicago <http://www.travelinfo.org/GCM/garychicago-java.html>
URL 12.19 Bay Area Traffic Report <http://www.bayinsider.com/autos/traffic/>
URL 12.20 Current CompuTraffic Report
               <http://www.computraffic.com/traffic.phtml?city=St.+Louis>

web technology. So it is reasonably safe to assume that there will be rapid growth in the use of interactive maps, both static and dynamic.

Figure 1.3 in Chapter 1 has the provider playing a central role. This role has been confirmed by most of the later chapters, although there are already indications, expressed clearly in Chapter 2, that the provider of the future may not simply be a provider of *maps* but instead of *data* and the *tools* with which the user can create maps from selected data. We will return to this topic later.

One fact stands out very clearly in the section of the book dealing with web map use and users, and that is the very uneven present worldwide distribution of access to the Internet. It is also clear that this situation is changing so rapidly that any estimate is out of date the day it is published. The Internet is really conquering the world with unbelievable momentum, far exceeding the speed of the personal computer revolution that has made this new revolution possible. Enormously wealthy new companies are mushrooming, and some of them are very interested in maps. This situation is bound to lead to increasing competition for the attention of the growing army of users, with of course more profits for the winners. "Give the users what they want" will be the watchword, and if the users demand better Internet maps, for example, better maps they will get. Urgent questions that will require answering include payment for the downloading of maps and geographic data, copyright, privacy and security.

Chapter 5 has shown that many of the basic "traditional" cartographic principles still remain valid in the era of web cartography, but of course augmented by many new capabilities introduced by the new medium. Some of these new capabilities are described in Chapter 7 and many more will undoubtedly become available as technological developments continue apace. The advent of a vector graphics web standard is an example, also the fact that software will appear on the market which will make the setting up of websites and putting maps into them very much simpler than they are now. The next technological "sub-revolution", bringing the power of the Internet to Personal Digital Assistants, will increase the demand for web maps and at the same time force providers to present all Internet data, including maps, much more efficiently than is done at present.

Despite the optimism of the outlook, a frequent remark made by our fellow authors as they ploughed through hundreds of websites in search of cartographic gems for the application chapters was, "Well, more hours gone and not much to show for it." Indeed, many sites, even those which make considerable use of maps, are disappointing from the cartographic design point of view and in the limited use they make of the technical capabilities of the Web. However, on paying a return visit to the sites definite progress could sometimes be observed, so perhaps the optimism is not entirely misplaced.

So far, we have considered the outlook on the basis of conclusions reached in the earlier chapters, based mainly on the extrapolation of current trends. For the longer term we are entering uncharted waters, and forecasting becomes very uncertain. We are, however, reasonably safe in assuming that, although static (interactive) maps for presentation purposes will remain, the need for dynamic maps will increase because of the demand for a more exploratory environment.

From a technical perspective progress is impressive. Many recent and future developments can be followed at the site of the World Wide Web consortium

(W3C, *URL 13.1*). For instance, developments such as those related to the Scalable Vector Graphics are worth keeping an eye on. Another important initiative is the OpenGIS Consortium (*URL 13.2*), that is working towards a more open and flexible use of different GIS software and data. The OpenGIS consortium has remarked on the importance of the Web for dissemination of geospatial data but has also noticed its limitations. In the wider context of the geospatial data handling process problems observed by the OpenGIS consortium include the limited ability to integrate geospatial data from different sources and to share geoprocessing resources. From this perspective the consortium started the Special Interest Group Web Mapping Testbed. Its aim is to create open geoprocessing web technology and to offer specifications as well (*URL 13.3*). Experiments are currently in progress. One example of how the future of web mapping according to the OpenGIS consortium will look like can be found at *URL 13.4*. It deals with disaster mapping in the southeast of the United States.

The new opportunities offered when the OpenGIS objectives are realised will not mean the end of developments. These opportunities will actually further increase the demands. An example is collaborative exploration and information visualisations. In this case not only the information is gathered from different locations but the users working on the same project will also be at different locations. Users will access and manipulate the same information while being able to communicate among each other in a virtual world (see Brutzman, 1997)). This requires what has been described as Internetworked graphics (Rhyne *et al.*, 1997), with at present some concentration on three-dimensional visualisations using VRML. The use of three-dimensional geospatial data on the Web is currently limited to relatively small data sets. This will also change and many experiments with geospatial data and VRML are ongoing (Lin *et al.*, 1999). The VRML-standard is being expanded into GeoVRML (*URL 13.5*). This extension will incorporate functions, obvious for those active in geosciences but less known in the computer graphics world, dealing with coordinate systems, terrain representations and level of detail (generalisation).

Another exploratory application that demands a sophisticated environment and advanced web mapping techniques deals with visual geospatial data mining. This refers to the extraction of implicit knowledge, geospatial relationships, or other patterns not explicitly stored in geospatial databases (MacEachren *et al.*, 1999). This approach brings together many advanced linked interactive visualisation techniques such as animation, three-dimensional representation and all kinds of diagrams and graphs (Wong, 1999). Similar demands might be raised when mapping cyberspace, since many of the graphic representations used deal with the third dimension and time series.

A topic not touched upon in this book is the role of web maps in education. The Web offers many opportunities to support education. Individual lectures, course modules or even whole courses can be offered via the Web (Cartwright *et al.*, 1998). For instance this book can be used in courses on Web Cartography and is supplemented by a website (*URL 13.6*). Other examples of websites supplementing textbooks can be found at *URL 13.7* & *URL 13.8*. Returning to the level of a single course topic the Web can support it with additional illustrations of the maps or any other geo-problem discussed in the course. Materials used during traditional lectures normally go back on the shelves, but via the Web they remain accessible. Additionally links to other examples can be given and the maps can be interactive and in full colour (Myers, 2000). The impact of the Web in education is not limited to the local university or college computer network. It can be made accessible to all. This calls for distance learning and virtual campuses (*URLs 13.9 & 13.10*). A different use of the Web and web maps in education is as preparation and (partly) replacement of fieldwork in for instance geography and geology (Moore *et al.*, 1999) (*URL 13.11*). Other interesting web map applications deal with public participation (*URL 13.12*, Krygier, 1999). People can discuss with their (local) governments on for instance new projects in a planning phase, or respond to other local policies.

Are all the frenzied activities around the World Wide Web and Cyberspace just a "hype" or are they really important? We would guess that the question has been answered by this book, certainly as far as cartography is concerned. The World Wide Web will be the medium of the future to work with geospatial data and to publish maps. Probably there will still be limitations as seen from different perspectives but these will be solved and as can be expected replaced by new limitations because of new demands, which in their turn will be solved. However, the World Wide Web has given and will continue to give the cartographic discipline new and fresh impulses. Soon the new opportunities offered, such as dynamic interactive multi-dimensional maps will be understood and incorporated into the existing cartographic theories.

URLs

URL 13.1   W3C consortium <http://www.w3.org/Graphics/SVG>
URL 13.2   OpenGIS <http://www.opengis.org>
URL 13.3   OpenGIS WMT <http://www.opengis.org/wmt/index.htm>
URL 13.4   WMT-example <http://homer.socialchange.net.au/webmap/ogcwmt/demo>
URL 13.5   GeoVRML <http://www.geovrml.org>
URL 13.6   Web Cartography < http://kartoweb.itc.nl/webcartography/webbook>
URL 13.7   Education <http://www.deasy.psu.edu/index.html>
URL 13.8   Book support <http://www.prenhall.com/knox>
URL 13.9   Distance learning < http://www.unigis.org>
URL 13.10 Virtual geography
              <http://www.utexas.edu/depts/grg/virtdept/contents.html>
URL 13.11 Virtual field course < http://www.geog.le.ac.uk/vfc>
URL 13.12 Public participation
              <http://www.owu.edu/~jbkrygie/krygier_html/lws/chang.html>

REFERENCES

Brutzman, D., 1997, Graphic Internetworking: bottlenecks and breakthroughs. In *Digital Illusions*, edited by Dodsworth, C., (Reading, MA: Addison-Wesley), pp. 61-97.
Cartwright, W., Fraser, D. and Pupedis, G., 1998, Hypereducation: prospects for

delivering region-wide cartographic science programmes on the web. *Cartography*, **27,** (2), pp. 27-40.

Krygier, J., 1999, Wide Web mapping and GIS: an application for public participation. *Cartographic Perspectives*, **33,** pp. 66-67.

Lin, H., Gong, J. and Wang, F., 1999, Web-based three-dimensional geo-referenced visualization. *Computers & Geosciences*, **25,** (10), pp. 1177-1185.

MacEachren, A., Wachowicz, M., Edsall, R. and Haug, D., 1999, Constructing knowledge from multivariate spatiotemporal data: Integrating geographic visualization with knowledge discovery in database methods. *International Journal of Geographic Information Sciences*, **13,** (4), pp. 311-334.

Moore, K., Dykes, J. and Wood, J., 1999, Using Java to interact with geo-referenced VRML within a virtual field course. *Computers & Geosciences*, **25,** (10), pp. 1125-1136.

Myers, J. D., 2000, Teaching with a web gallery. *Geotimes*, (1), p. 26.

Rhyne, T. M., Brutzman, D. and Macedonia, M., 1997, Internetworked graphics and the web. *Computer*, (8), pp. 99-101.

Wong, P. C., 1999, Visual data mining. *IEEE Computer graphics and applications*, **19,** (5), pp. 20-21.

APPENDIX A

File formats and plugins

Wim Feringa

A.1 INTRODUCTION

Standard HTML has very limited capabilities for publishing geographical data on the Web. Many extras have to be added both on the server side as well as on the client side. The publisher of the data needs a clear idea of the functionality and content of what is published and has to find the correct ways to visualise it. To publish maps requires putting them through several different software packages. In addition, publishing on the Web demands specific file formats, a special colour space, a web-oriented design and a carefully considered navigation system. Data (images, maps, complete designs) publishers want the user to be able to view and possibly interact with them in the manner intended. This will entail that the viewer of the data also needs to install extra software to be able to look at the information offered over the Web. In this Appendix an overview is given of many file formats and plugins relevant to publishers and users. Basic requirements and concepts have already been introduced in Chapter 6.

The WWW is monitored and described by the World Wide Web Consortium (W3C) (*URL A.1*). This consortium defines the standards for HTML (HyperText Markup Language) and the different (graphic) formats supported. Browser developers (like Netscape and Microsoft) however often make their own additions or small changes to that standard. This can result in data, prepared for one browser, becoming not correctly displayed or not displayed at all in another one. Commercial companies themselves also introduce new graphic and multimedia file formats, which they try to set as a new standard. For these new formats additional software components (the so-called plugins) have to be downloaded and attached to the browser in order to be able to view them (see Figure 6.5). Not only graphic file formats have to be considered, but also the extra facilities the WWW has to offer, such as animation, sound and video.

The URLs and examples given in this Appendix do not necessarily refer only to cartographically related products, since the intention is to give an overview of the techniques in general. The technological aspect of the WWW is changing rapidly and new developments are reported in the website related to this book.

A.2 CODING/AUTHORING

When the World Wide Web began there was only the possibility of transferring simple HTML coded pages, with text as the only content. During the years users became more demanding, wanting to be able to format the text and add images to it. Today the basis is still the HTML coding, but many additions or applications have been written to enable image and data transfer over the Internet. The overview given below is not complete, but it gives an idea of what can be done and what is still needed.

HTML

The basis for all Internet communication is the coding of the information that has to be displayed on the user side. The browser, in which the coded files are loaded and displayed, understands predefined codes. The files needed for display in a browser are written in HTML, which describes the content of a web page. Such a HTML file can be produced in a simple text editor, and is in itself not spectacular. All information for the browser is written between these < (opening) and > (closing) symbols, called "tags". Such a tag is called a "Markup" indicator and informs the browser how the file should look when it is displayed. For example, <BR> indicates a line break and forces text to start on a new line.

In itself this "Markup Language" is hardly responsible for the big boom of information transfer over the Internet. Without the "Hypertext" glued to it, only boring, non-interactive pages would be the result. The strength of the Web is the ease of "jumping" from one place to another, made possible by the concept of hypertext. Information can be organised by creating a possibility to link one information unit to connected associations (named a hypertext link), to jump from one part in the file to another or to completely different pages or even different sites. The user can choose how and in what sequence to receive the information. With the invention of the hypertext concept (by Ted Nelson, for development of the Xanadu system) the step to the World Wide Web was quickly made (*URLs A.2 – A.4*).

A new development is Dynamic HTML, a term for a combination of new HTML tags and options (layers, time lines, etc.), style sheets, and programming. The idea is to create more options for animation and to make a website more responsive to the user's action. The game in *URL A.5* demonstrates possible interaction. Part of the Dynamic HTML is specified in HTML 4.0 (*URL A.6*), but here a difference in behaviour of the two main browsers is noticeable. HTML 4.0 is supported by the latest versions of both the Netscape and Microsoft browsers, but some additional dynamic capabilities are only supported by one of the two. Another problem in using Dynamic HTML is that many users still have older versions of the browsers, and therefore not everyone is able to use the functionality offered by it or even may not be able to view it at all. A solution is to create two different sites, to serve everyone. For an introduction to HTML refer to *URL A.7*. See also Section 6.3.1, on basic web data formats.

Java

Java is a platform-independent programming language that can be used to build complete applications that may either be run on a single computer or within a network. The small applications, also named applets, can be used as part of a web page. These applets make it possible for a web page user to interact with the page. In the case of maps, the user can interact with the map data. To be able to view these Java applications, the "Java virtual machine" is required (standard part of the latest version browsers), which will run the applet. There are many Java-based maps to be found (see *URL A.8* and *URL A.9)*. See also Section 6.4.1.

JavaScript

By using only HTML, a direct response to any user action, other than opening a new page or another location, is not possible. To create more interactivity, JavaScript code can be added to the HTML code. Initially this was designed (by Netscape) to control Java applets, but now the possibilities reach far beyond that. JavaScript nowadays is used in websites for example for creating an "on-mouse-over" effect, where one image is swapped for another. The script allows response to web-based forms and direct processing of the data. It can be used to create popup windows, to change page elements on the fly, to create and store data on the user's machine, to change dates on a web page, etc. (see *URLs A.10 – A.12)*. See also Section 6.4.1.

WAP

The Wireless Application Protocol (WAP) is an evolving standardised specification for wireless Internet access (mobile phone, pagers, Personal Digital Assistants (PDA) and others), initiated and controlled by the WAP forum which was started by Ericsson, Motorola, Nokia and Unwired Planet (now Phone.com) (*URL A.13)*. The WAP forum is an industry group and has now over 100 members, so the forum can be compared to the W3 consortium (*URL A.1)*. The specifications follow closely the architecture of the Internet and the World Wide Web since this has proven to be stable and successful. Within the WAP, protocols, languages, hyper linking, scripts and all kinds of formats show similarity to the existing Internet technologies. The whole set of protocols and scripts is optimised for use in a wireless environment.

The size of the display is very limited, sometimes to only one line. The display unit is not defined as a page, where the user can scroll to get more information, but as a "card" (one screen of information). A card can contain text, images, links and input fields. For a set of information it can be that more cards are needed, creating what is called a "deck". When calling information, first the whole deck is loaded before a card can be displayed.

The WAP is structured in three main layers, the application layer (WAE), the session layer (WSL) and the transport layer (WTP). The Wireless Application Environment (WAE) contains amongst others the Wireless Markup Language (WML) (based on HTML) and WMLScript (based on JavaScript). In the Wireless Session Layer (WSL) provisions are established for connection with the application layer. In this layer are to be found the Wireless Session Protocol (WSP) (to be compared with HTTP) and Wireless Transport Layer Security (WTLS) for encryption of all session data. Wireless Transport Protocol (WTP) belongs to the third transport layer and can be compared with the TCP for the Internet.

Small portable devices have small, monochrome and rather low-resolution displays, so that the content has to be very basic, with simple graphics and few elements per screen. This of course also applies to maps. See for more information on this subject *URLs A.14* and *A.15*.

XML

In 1998 the W3C approved XML (eXtensible Markup Language) version 1.0 specifications. XML is more or less a subset of SGML (Standard Generalised Markup Language), the international standard for defining descriptions of the structure and content of different types of electronic documents.

Web developers had felt the need for a better way of putting data on the Web. They wanted a language with a better structure, better organisation, wider support and more flexibility than HTML. HTML describes the content of a web page (text and images) only in terms of how it is displayed and how to interact with it. In between the <body> </body> tags there is complete chaos as to how data are offered. Whenever there is a need for reorganising the data structure the entire HTML file usually has to be re-written.

XML describes the content in terms of what is being described. As with HTML tags are needed but the tags can be completely customised, meaning that one can create one's own way of structuring the data (extensibility). It is also possible to use conventions in organising data and to use common structures, so the data can be shared. Unlike in HTML every markup has a meaning; it gives information about the data described between the tags. <NAME>Wim Feringa</NAME> is a valid tag, giving a meaning to the data "Wim Feringa".

XML files are meaningless when not accompanied by a DTD file (Document Type Definition). The DTD file handles the definition of the valid markup, it provides applications (not necessarily Web browsers) with advance notice of what names and structures can be used in a particular document type. This DTD takes care that all documents belonging to a particular type will be constructed and named in the same way. DTDs can be shared or even promoted to a "convention", so that specialised user groups (e.g. geoinformation group) are able to structure data in a more general way, making it easier to exchange information or to obtain the required data.

A third file, a stylesheet, is needed to be able to publish the data on the Web. This stylesheet gives instructions to an application (like a web browser) on how to display the information. Such a stylesheet can be written in a number of style languages, such as XSL (eXtensible Stylesheet Language) or CSS (Cascading Style Sheet).

Finally, a processing program is needed to bring together the XML, the DTD and the XSL or CSS together in a meaningful document. XSL is integrated into the Microsoft XML processor that is part of Internet Explorer 5. It transforms XML into HTML, which is then displayed using CSS. Developments are speeding ahead and the expectations are that it will not take too long before others will follow. In the end XML will probably replace HTML and by then most likely a set of tools will be offered to generate a whole set of XML files in the same way as the HTML editors do the coding now for the web designers. For working examples see *URLs A.16* and *A.17,* using Microsoft Internet Explorer to see them functioning. An extensive overview of this subject can be found on the SGML/XML web page (*URL A.18*). Also see *URLs A*.19 and Connolly (1999) and Garshol (1999).

A.3 GRAPHICS

When discussing implementing maps on the Web, graphics have to be included. There basically is no limit in the web specifications to graphic formats that can be used on the Web, but note that many need a suitable plugin or a special viewer. Without this, the choice of usable formats is limited. The file formats supported as standard are described by the W3C. When using one of these one can assume that all users will be able to see the images in the browser. Besides the file format limitation, the file size (in Kb) is of great importance, knowing that larger files take longer to download. Unless they are certain that they must have the information on a particular Internet page, most users tend not to wait too long: after 10 seconds they get impatient and after 20 seconds they are gone. Therefore, an image and all other information on a page should display within seconds. At present there are two widely used standard graphic formats (GIF and JPEG), both raster-based. Vector-based formats are not yet fully supported and plugins are needed to view them.

GIF

On the Web the GIF (Graphics Interchange Format) has become standard. This format for compressed files is actually owned by Compuserve. At the start of development of putting images on Internet pages in 1987 the relatively low data transmission rates formed a problem. Even nowadays any saving in transmission time by compressing the data is worth pursuing. The GIF compression is "lossless", meaning that after decompression the image looks the same as before compression. It uses a LZW (Lempel Ziv Welch) (*URL A.20*) compression, making the file size very small. LZW is a way of compressing data that takes advantage of the repetition of strings in the data. Raster data can contain a lot of this repetition, so LZW is a good way of compressing and decompressing them. A GIF file consists of a header and the encoded data. The browser uses the LZW scheme in reverse to decompress the file into a bitmap for display.

Figure A.1 A sequence of GIF images needed for an animation.

In 1989, Compuserve added the "interlaced" and the "transparent" features as well as "animated GIF". With interlacing an image displays with a gradual increase in the resolution, so that an initial low-resolution but complete image will show very quickly during download. Transparency is a very important feature for web designers. A specific (indexed) colour can be indicated as transparent. Nonsquare shapes can now show nicely on top of a background. A GIF animation is nothing more than a sequence of GIF images (Figure A.1) played at a pre-set speed. Siegel (1996 and *URL A.21* refers to it as a "poor-man's video".

GIF compression is best used for images that have solid colours. Most (thematic) maps are within this category. GIF is also a good format to use when a simple animation has to be made. Many raster editing software packages support a direct export to this GIF format. A warning note here is that GIF supports only 8-bit colour, giving a maximum of 256 different colours in a single image. For images that have graduated colours, as for example a photo image, another compression technique, JPEG, is advisable.

JPEG

The name is derived from the committee that designed it, the Joint Photographic Experts Group (*URL A.22*). The group was formed especially to come up with a good method for compression of photographic and photo-realistic images, with a colour depth of 24 bits. JPEG provides true colour and compresses image files far more than the GIF compression method. The disadvantage is that it uses a "lossy" compression method with the result that part of the image data will not be recovered when decompressing.

The JPEG compression is based on the fact that the human eye is far less sensitive to small changes in the hue and saturation aspects of colour than it is to small changes in brightness (lightness). The method therefore compresses the hue and saturation information much more than the brightness information. The amount of data reduction depends on the compression quality settings. Note that since the header information consumes a good deal of space, JPEG does not produce smaller file sizes than GIF (maybe even larger) for small images of less than 100x100 pixels. For these sizes it is better to use the GIF compression technique. The types of map or image for which you would use JPEG are those with graduated tints (e.g. hill shading) and/or photographs. For more information on JPEG see *URL A.23*.

PDF

A PDF file (Portable Document Format) of Adobe is a copy of an original application file, converted to a (third version) PostScript file format, that can be viewed in a browser by using the Adobe Acrobat Reader plugin, which can be downloaded free of change. This will run on a wide variety of platforms, such as DOS, Windows, Macintosh, and UNIX. A PDF file contains a view file that shows the page as created. Furthermore, it can contain embedded type, graphic objects (bitmaps and vector images), hypertext links (internal as well as external), links to external files such as sound, QuickTime or AVI movies and links for variable forms data. The PDF format is highly suited for distributing maps through the Internet. A PDF file of a map can be generated, with the aid of Adobe Acrobat Distiller, directly from within the application where the map is produced. When generating the portable document the compression rate for images is selected. For viewing in a browser, purely for screen display, a high compression rate can be selected, resulting in a very small file size. If one wants the user to be able to print a high quality image one should select a low compression rate, remembering that the resulting larger file will take longer to download. An advantage of PDF is that all kinds of security options can be added, such as password protection for opening or non-printing and non-saving options. The Acrobat Reader plugin gives the option of zooming in or out and panning. Using PDF files, an interactive brochure or map can be made with active links and a fully preserved layout and good image quality, ready for publication on the Internet without the need for specific coding. For more information see *URL A.24*.

PNG

One format that so far has not received the attention it deserves is the PNG (Portable Network Graphic) format. If one considers its advantages (no more differences in colour display on different platforms because of colour control through so-called gamma settings; real transparency by specifying an alpha channel so that a transparent layer really blends into whatever colour is underneath; smaller files than similar GIF images; always editable and possibly to resave without loss of information; royalty-free) one wonders what went wrong. Why is this format not as popular as the GIF format?

One of the reasons is the fact that not many producers of web pages know about the existence of the format. The most important reason however is the lack of browser support and browser compatibility. The latest browser versions support the native PNG format, but not fully (gamma functionality still is not included) and they tend to give errors when opening a PNG file. For those who publish maps on the Web this format offers some extra advantages such as more or less protected files (they cannot be saved to the hard disk) and the possibility of editing vectors and text in the image using software such as Macromedia Fireworks or Adobe ImageReady. For more information see *URLs A.25* and *A.26*.

SVG

Almost all line images seen on the Web are in fact raster images, either GIF or JPEG. Raster images have some disadvantages: they are scale dependent and every pixel has to be described. It would be much better if these images could be described as line images, i.e. as vectors and fills. The W3C is working on it and they have produced a format with the name SVG (Scalable Vector Graphics). It is a graphics format written in XML and styleable with CSS (Cascading Style Sheet). The consortium itself expects this format to become a popular choice for including graphics in XML documents.

A vector description of an image consumes much less memory space than the equivalent raster description. For a line only the description of a few points are needed including the specification of thickness and colour. For an area a description of the fill is included. The good thing about vectors is that the graphics can be scaled (up or down) without sacrificing quality.

SVG is a purely text-based collection of XML-like commands, generating graphics that require no plugins or extra tools to be visible on a website. It can fully integrate with other text, graphics and commands in the file, even with HTML. SVG allows three types of graphic objects: vector graphics, images (raster) and text. The drawings can be dynamic (animation along a predefined time path) and interactive (create on-mouse-over, on-click, etc. and link to other parts or other areas). Further it supports many features that can now be used in raster editing packages (like clipping, filter effects, masking, gradients, opacity, transformations) but also good colour management and output options. Elements can be placed on a page accurately and fonts can be used with much more freedom. And since it is all text based, download times are really limited.

Considering the authors of the specification it is likely that SVG will be widely supported. The companies that have a member in the developers' group are (among others): Microsoft, Adobe, Sun, IBM, Netscape, Apple, Macromedia, HP, Corel and Quark. It is to be expected that SVG will be supported and can be used as from browser version 6.0. More information on this topic can be found at the website of the W3C (*URL A.27*) and, among others, at the site of Adobe (*URL A.28*).

A.4 ANIMATION

When entering certain sites, one may be dazzled by all the moving and flying elements. Mostly these are GIF animations, a sequence of GIF images in a fixed order seen from a fixed position, not influenced by the user. They can be used, for example, for showing development of an area or growth of a city, i.e. simple straightforward timelines without options of sidestepping or interference by the user (*URL A.29*). For a more advanced visualisations the file types described below give great possibilities for publication of maps on the Internet.

SWF

More or less everything that SVG promises to do is possible with the SWF (Shock Wave File) format of Macromedia. However, for a SWF special software is needed in order to be able to view it, the so-called Shockwave & Flash player. The Shockwave file is created in Macromedia Director and is a real multimedia product with interaction possibilities. It is much used for games and multimedia productions on CD-ROM. A flash file is created in Macromedia Flash and is meant for creating web pages with highly sophisticated interactive interfaces, animation and information graphics. The players are supported as standard by the latest versions of the Netscape and Internet Explorer browsers. It is not really unexpected that Macromedia has taken a place in the developers' group for the SVG. More information can be found on the website of Macromedia (*URL A.30*).

An increasing number of sites can be found containing Flash animations. Compactness and scaleability is what makes the format so popular, as well as the possibility of adding sound and video as well as interactivity. Another element supported in a Shockwave file is audio, so one can add all kinds of sound to the animation. A nice example of a Shockwave file, including a map, is a game where photographs have to be taken according to a certain assignment (*URL A.31*). An example of a website that uses a kind of map and is created with Flash is the opening page of the Division of Geoinformatics, Cartography and Visualisation of the ITC (*URL A.32*). For another example see an Internet store site at *URL A.33*.

VRML

VRML (Virtual Reality Modelling Language) offers the experience of "walking" or "flying" around in a town or building, making one's own decisions on which way to go. It is a language for describing three-dimensional (3D) space, for distribution over the WWW. The 3D scenes can be viewed by using a VRML browser, mostly working together with a HTML browser. Details of the specifications can be found at the website of the W3DC (*URL A.34*). For a primer and tutorial see *URL A.35*.

The 3D elements (also called objects) can be produced in different ways. They can either be created as an ASCII text file (considerable programming knowledge is needed for this) or converted into VRML from commonly used 3D file formats by the use of an automatic translation program. A VRML object is scaleable and one can zoom in or out, rotate, turn around, pan, etc. An object is defined by points in a 3D (xyz) space. Connecting the points results in a "wireframe" on which all kinds of surfaces can be wrapped. The surface can be rendered with a colour and structure and a varying amount of reflection or transparency can be added. Settings can also be given for how an object is illuminated, including the strength and direction of the light. An object itself can also emit light, influencing other objects. So-called VRML worlds are constructed out of 3D objects (not necessarily defined in VRML), placed in a 3D environment. The detail and quality of the objects depend of course on the settings for rendering an object. VRML files can show a surface only as a polygon or as a set of polygons, depending on the complexity of that specific object (e.g. a globe is rendered as a collection of triangles).

VRML viewers must be compatible with the files: files produced for a later version of VRML cannot be viewed on an early version viewer. The newer viewers allow several rendering options: wireframe (only the lines are shown, very fast display); shading per object (flat) gives a very rough display but is fast; more detailed shading (Gouraud) where every polygon gets its own shading (slower than the flat shading); realistic shading (Phong) where the shaded elements are very small and polygons are rounded off (very slow rendering); either show or do not show the texture.

VRML files are basically ASCII files, and for that reason they can be very small. However, if the producer includes all kinds of textures and surfaces the file size can become too large to download in a reasonable time. An example of a single building in VRML is the building of the ITC in Enschede (*URL A.36*). Two more complex examples of a VRML world are Schiphol Airport (*URL A.37*) and Red Square in Moscow (*URL A.38*).

A.5 VIDEO

Presenting maps over the Internet offers great opportunities for the use of different media, so why stick to a flat and lifeless paper-like representation of the data? VRML offers the possibility of creating a 3D world. Video and linked VR can add yet another level of information to web maps and change them to a multimedia experience. The problem with video images for the Internet is to achieve a good balance between quality and file size. Better compression techniques as well as the increasing capacity of the connecting hardware are improving the situation to the point where video is now a reasonable option.

For production of video for the Web special video editing software is required that is able to save the video in the AVI format. Pieces of film have to be edited to make a sensible story, titles have to be added as well as a pleasant background sound. Take care that videos intended for use on the Web tell a clear story, are short in time and small in terms of bytes, but still of reasonable quality. The latest development is "streaming video", where live broadcasting is possible. This can be useful for example for monitoring stretches of road liable to traffic jams (*URL A.39*). This technological development is more demanding on the hardware. The large multimedia files involved are stored on servers with a very high performance. Bandwidth and maximum serving capacity are crucial for the overall performance and accessibility of the site and of the data being served.

AVI

A file format for digital video and audio for Windows, defined by Microsoft and named AVI (Audio Video Interleave), is one of the most popular formats for presenting video over the Internet. The AVI format is already an "old" format, developed before the real take off of video on the Internet. The main purpose is for storing and playing back digital video from a CD-ROM or a local hard disk. AVI can be compared to QuickTime (see below), since they both use similar techniques and aim for the same type of use.

The file format is cross-platform compatible, allowing AVI video files to be played under different operating systems, if of course the proper player is installed. When producing an AVI file one can make a choice from a wide range of video quality parameters: image resolution, frame rates (it is recommended not to take less than 15 frames per second), colour encoding (256 to millions) and sound (5Khz mono to CD quality). Speed of data transfer is most important for the Web. Therefore fairly low quality settings may be best for this purpose (see McGowan, 2000).

MPEG

Like JPEG the name MPEG is derived from the name of the committee that worked out the specifications for this standard. The full name of the committee is the Moving Picture (Coding) Experts Group (*URL A.40*), founded in 1988 under the auspices of the International Standards Organisation (ISO). The group had to come up with standards for digital audio and video compression. Results so far are MPEG-1 (1992, standard for storage and retrieval of moving pictures and audio), MPEG-2 (1994, standard for digital television), MPEG-4 (1998, standard for multimedia applications also for the WWW) and MPEG-7 (to be expected in 2001 as the content representation standard for multimedia information search, filtering, management and processing). A MPEG viewer or player is needed which can be downloaded from different places, and will run on different platforms. Being an ISO standard ensures that this format is widely supported.

QuickTime

QuickTime is a technology developed by Apple and was one of the first video formats for the PC (*URL A.41*). The older versions of the QuickTime format are quite similar to the AVI format, while the latest version can be compared more to the ASF format of Microsoft (see Section A.7). Similar techniques for compression and decompression are used. At first the QuickTime format was supported only by the Apple Macintosh computers, now it is a format that is supported on all platforms. QuickTime is a multimedia development, i.e. storage and playback technology and sound, text, animation and video are combined in a single file. QuickTime supports a very wide range of file formats, video compressors, video effects, sound compressors and web browsers. The latest version QuickTime 4 also supports panoramic views, better known as VR movies. These panoramic views can be generated with special tools that can stitch a series of digital photographs together into one. Many examples of these files can be viewed at Apple's showcase site (*URL A. 42*).

A.6 SOUND

Sound is not really one of the options one thinks of when discussing elements that could be part of a map. Sound is often an integrated part of video files linked to a map, but sound as a stand-alone element can provide an extra level of information. It could for instance give the correct pronunciation of place names, or provide extra information about resorts in a tourist map. See *URL A.43*.

Audio formats can be classified in several ways: RAM-based (the file is loaded into the RAM before playing) disk based (the file is first loaded on to the hard disk before playing); self-describing (the header contains options for encoding and possibly other information) or raw (the encoding is fixed); self-contained (no extra server support is required) or streaming (requires a special server for processing). This last classification is most applicable to distribution of audio files over the Internet. As with video, sound can be "streamed", meaning that after downloading a few large sections of the whole file, the software can start playing the sound. In this way large files can be used, without causing frustration to waiting users. However, note that as with video, problems can arise related to server performance and bandwidth, especially with streaming audio.

There are many sound formats that can be used on the Web, but it makes sense to use a format that is widely supported. Before the user can listen to a sound file sometimes first a plugin or separate helper application has to be downloaded (often sound and video are combined in such an application). For sound editing special software is required. For standard tasks there is much freeware and cheap shareware available on the WWW while professional software can be purchased for more demanding editing jobs.

AIFF

The sound format developed by Apple for use in the Macintosh Operating System is AIFF (Audio Interchange File Format). The file contains information about the number of channels (mono or stereo), information for applications, bit depth (the more bits used, the more accurate the resulting output will be) and sample rate (the number of samples of a sound per second). With high quality sampling and maximum settings, the output will be the same as the original, resulting in a very large file size. For CD-ROM productions this is not a problem, but for distribution over the Internet it is better to use lower quality settings to keep the file sizes small. The format is widely supported by many applications on all platforms.

AU

The AU (audio) format originates from SUN and is designed for UNIX operating systems. The Internet is still dominated by UNIX-based servers, and so AU is much used. Most browsers support the AU file format by default. This ensures that in principle it is the most widely supported and therefore safest to use. The AU files are relatively small because of the compression method and they can therefore be downloaded rather quickly. What works against this format is the poor output quality, so it can be best used for sound effects like clicks and gongs and for speech.

MIDI

For the use of background music (wallpaper sound) the MIDI (Musical Instrument Digital Interface) format is very suitable. This format does not reproduce natural sounds, it synthesises sounds, so that, in contrast to the other formats, the music itself is not recorded and no sound is stored. A MIDI file is a set of instructions about the notes that should be played and about the order, the timing, the volume and the type of instrument they should be played on. The sound that is played depends very much on the type of sound card in the PC. The latest versions of the browsers support MIDI files and no extra software is required.

MP3

This format is defined by the MPEG commission (see the Section on video MPEG) and the name is an abbreviation of "MPEG-1 Audio Layer-3". The MP3 compression of the file is very high compared to other formats, about $1/12^{th}$ of the size while preserving the high (CD) quality when played. MP3 uses a compression technique that removes those parts of the sound that most listeners cannot hear. In addition, it can include advanced technology to regulate the use of the file. For example, one can add a digital watermark, prevent files from being saved and/or copied and restrict playback to only one PC. Streaming MP3 is possible and it could be the base format for live Internet radio, but at the present time the files are usually downloaded and played later locally.

RA

A very common and widely used Internet audio format is RA (Real Audio), which supports also streaming audio. The format requires a special type of server that is capable of supplying the information in the correct way. The RA format can be very compact, but if very high compression rates are used the disadvantage is lower quality output. The software to play these RA files is freely downloadable and is multi-platform.

WAV

Microsoft developed WAV (Wave Form Audio File Format), which is the native sound format for Windows. The format is similar to the AIFF format and has more or less the same content. The WAV file format contains uncompressed raw audio data, information about the number of tracks used (mono or stereo), bit depth and sample rate. Because of its origins it will be obvious that this format is widely supported on all platforms.

A.7 SYNCHRONISED MULTIMEDIA

As stated earlier, the Internet allows the presentation of information by making use of a combination of media, such as text, images, sound and video. Until recently, web developers were always confronted with the impossibility of controlling how their data were presented at the user's end. They also could not predict the effect of the user's particular configuration. All the separate elements of a HTML page are different in size but ideally they should be downloaded all at the same time, which is not possible. This is not to mention the headache the web designers had to suffer when trying to make a perfectly timed show, where one image should follow after a fixed number of seconds with perhaps half way a pleasant background tune coming in.

A rather new development in presenting multimedia on the Web is to synchronise text, images, audio and even video (Bouthillier, 2000). To achieve this two approaches are possible: use a specially designed mark-up language developed by the W3C (named SMIL (Synchronised Multimedia Integration Language)) or use a protocol that supports streaming developed under the supervision of IETF (Internet Engineering Task Force) and named RTSP (Real Time Streaming Protocol). The IETF is an "open international community of network designers, operators, vendors, and researchers concerned with the evolution of the Internet Architecture and the smooth operation of the Internet" (*URL A.44*).

An important element of the properly synchronised method of presenting media to the user is the support of streaming. As has been mentioned already, this is the technology that makes it possible to view the data during downloading (transmission) over the network in real time. There are of course many technical problems to be solved to make it finally possible to ensure a trouble-free use of this technology. Sometimes regular web servers can do the job in delivering the data over the Web, but it is better to use specially configured servers. The reason is that the streaming data has to be delivered in "packets", which is the key to this technology. A packet is a block of data containing all the information needed for a certain part of the total presentation. The packets have to be delivered in a sequential order, otherwise the presentation would not make much sense or the player might refuse to show it. This technique makes it possible to present data on the Web as complete stand-alone multimedia shows including navigation, etc. This format is very suitable for training, education and, in cartography, for the presentation of multimedia atlases.

SMIL

To allow simple authoring of multimedia shows with possibly a need for controlled display of the elements, W3C has designed SMIL (*URLs A.45* and Hoschxa, 2000). RealNetworks and Netscape among others were involved in the working group. The SMIL language is a markup language like HTML, and it is easy to learn. This means that SMIL presentations can be written with a very basic text editor. A SMIL (pronounced as "smile") presentation can be composed of images, text or any other medium type including streaming audio and streaming video.

SMIL allows all these parts to be sent separately but it coordinates the timing. The media object used is accessed by its own URL which means that presentations can be made of objects stored at different places and objects can be reused also in other presentations. With SMIL, multiple language versions of soundtracks and videos can easily be retrieved, transparency and scaling can be defined and interaction can be built in.

Not every user has the fastest modem or the fastest connection. Using SMIL, these variations can be taken into account when data are shipped from server to user. Every single media object can be stored in a different resolution (or bandwidth). The producer makes more than one version of the objects (very often "low quality" and "high quality"). SMIL then automatically chooses the version that suits the user's particular configuration.

RTSP and RTP

HTTP (Hyper Text Transport Protocol) can collect the SMIL documents, but since HTTP is based on TCP (Transmission Control Protocol) that cannot cope with timelines, it is not really the correct protocol for handling multimedia sessions that include timelines. For this purpose the RTSP (Real Time Streaming Protocol) has been developed by RealNetworks, Netscape Communications and Columbia University (*URL A.46*). The whole development took place under the auspices of the IETF.

RTSP uses RTP (Real time Transport Protocol), which is a packet format for multimedia data streams. This protocol is also used by the ITU (International Telecommunications Union) for the two-way delivery of audio and video phone data, for example for video conferencing. RTSP is a multimedia presentation control protocol that takes care of an efficient delivery of streamed multimedia over IP (Internet Protocol) networks. The protocol ensures that the creation tools, the encoders, the servers and the players all use the same "language" so that they "understand" each other.

ASF

There are several file formats that support streaming and use the RTSP to deliver the data streams. One of them is the new Microsoft format ASF (Advanced Streaming Format) (Microsoft, 2000) that will in time replace the AVI and WAV formats completely. Unlike those two formats the ASF file supports streaming. ASF will also support local playback, as well as multiple language support, scaleable media types, etc. As with the AVI format it will be environment independent. But to be able to produce, to distribute and to view the ASF files special software is required.

Whereas SMIL indicates which file to load from where and when to play it ASF acts as a container into which all data for the multi media session are put. ASF controls the packets and organises their sequence.

URLs

URL A.1   The World Wide Web Consortium homepage <http://www.w3.org>

Coding
URL A.2   Xanadu <http://www.xanadu.com.au/general/>
URL A.3   Ted Nelson and Xanadu
            <http://jefferson.village.virginia.edu/elab/hfl0155.html>
URL A.4   Hypertext guru has new spin on old plans
            <http://www.wired.com/news/news/technology/story/11766.html>
URL A.5   Example of a DHTML-based page (only with Internet Explorer)
            <http://plaza.harmonix.ne.jp/~jimmeans/gymkhana/>
URL A.6   HTML 4.0 <http://www.w3.org/MarkUp/>
URL A.7   Introduction to HTML
            <http://kartoweb.itc.nl/webcartography/courses/basic>
URL A.8   A Java applet about projections
            <http://www.neosoft.com/~forge/java/Cartog/Cartog.html>
URL A.9   Collection of Java-powered maps
            <http://www.stjohnsprep.org/htdocs/sjp_lnks/maps_res.htm#interact-java>
URL A.10   Spotlight on JavaScript
            <http://builder.cnet.com/Programming/JsSpotlight/>
URL A.11   Jumpin' Java, the future of the Net
            <http://www.cnet.com/Content/Features/Techno/Java/>
URL A.12   How Java technology works with desktop computers running web
            browsers <http://www.sun.com/java/javaworks/desktopbrowser.jhtml>
URL A.13   The home page of the WAP Forum <http://www.wapforum.com/>
URL A.14   An info site on WAP <http://www.wapnet.com>
URL A.15   Another info site on WAP <http://www.wap.net>

URL A.16   XML: Demos and miscellaneous uncategorized
             <http://www.oasis-open.org/cover/xmlMisc1999.html>
URL A.17   A working Internet page based on XML
             <http://www.guy-murphy.easynet.co.uk/>
URL A.18   A comprehensive online database on the subject of XML
             <http://www.oasis-open.org/cover/xmlIntro.html>
URL A.19   A FAQ on the XML language <http://www.ucc.ie/xml>

Graphics
URL A.20   Basics of the LZW encoding scheme <http://www.geocities.com/
             ResearchTriangle/System/5245/programming/lzw.htm>
URL A.21   Website that accompanies Siegel, 1996 <http://www.killersites.com>
URL A.22   The official website of the JPEG committee <http://www.jpeg.org/>
URL A.23   FAQs about JPEG <http://www.faqs.org/faqs/jpeg-faq/>
URL A.24   Adobe Acrobat and PDF < http://www.adobe.com/products/acrobat/>
URL A.25   The PNG home page < http://www.libpng.org/pub/png/ >
URL A.26   The Story of PNG <http://www.sonic.net/~roelofs/test/story-of-png.html>
URL A.27   The W3C pages on Scalable Vector Graphics
             <http://www.w3.org/Graphics/SVG/>
URL A.28   Scalable Vector Graphics <http://www.adobe.com/svg/main.html>

Animation
URL A.29   GIF animation of the growth of Enschede
             <http://kartoweb.itc.nl/webcartography/webmaps/dynamic/dv-example4.htm>
URL A.30   The Macromedia website <http://www.macromedia.com>
URL A.31   A Shockwave example
             <http://www.photohunt.com/games.html>
URL A.32   The Division of Geoinformatics, Cartography and Visualisation of
             the ITC <http://www.itc.nl/carto>
URL A.33   Internet store Town 24 <http://www.town24.com/>
URL A.34   Web3D consortium, VRML specifications
             <http://www.web3d.org/fs_technicalinfo.htm>
URL A.35   VRML primer and tutorial <http://tecfa.unige.ch/guides/vrml/vrmlman/>
URL A.36   ITC building in 3D
             <http://kartoweb.itc.nl/webcartography/webmaps/dynamic/dv-example2.htm>
URL A.37   Schiphol Airport in 3D
             <http://www.schiphol.nl/engine/index_def.html?lang=en&page_nr=590>
URL A.38 Red Square in Moscow
             <http://www.parallelgraphics.com/scenes/redsquare.wrl>

Video
URL A.39   Live monitoring of traffic jams
             <http://www.accessarizona.com/news/traffic_cam.html>
URL A.40   The MPEG home page <http://drogo.cselt.stet.it/mpeg/>
URL A.41   About QuickTime <http://www.apple.com/quicktime/overview/>
URL A.42   Apple showcase <http://www.apple.com/quicktime/hotpicks>

Sound
URL A.43   Napoleon's 1812 Russia campaign.

<http://www.cs.cmu.edu/Groups/sage/sageshk.html>

Synchronised Multimedia
URL A.44   The   Internet   Engineering   Task   Force
<http://www.ietf.org/overview.html>
URL A.45   The W3C on the Synchronized Multimedia Integration Language
             <http://www.W3.org/AudioVideo/>
URL A.46   RealPlayer streaming audio and video
             <http://www.realnetworks.com/devzone/index.html>

REFERENCE

Bouthillier, L., 2000, What I did last summer
   <http://www.people.hbs.edu/lbouthillier/smil/> (accessed 05.07.2000)
Connolly, D., 2000, The XML revolution
   <http://helix.nature.com/webmatters/xml/xml.html> (accessed 05.07.2000)
Garshol, L.M., 2000, Introduction to XML <http://www.stud.ifi.uio.no/
   ~larsga/download/xml/xml_eng.html> (accessed 05.07.2000)
Hoschka, P., 2000, Toward synchronized multimedia on the Web
   <http://www.w3journal.com/6/s2.hoschka.html> (accessed 05.07.2000)
McGowan, J.F., 2000, AVI overview. Effective use of video on a webpage.
Microsoft, 2000, About ASF <http:www.Microsoft.com/asf/about.ASF.htm>
   (accessed 05.07.2000)
   <http://www.jmcgowan.com/avi.html> (accessed 05.07.2000)
Siegel, D., 1996, *Creating Killer Websites*, (Indianapolis: Hayden Books), p. 53.

APPENDIX B

Design, colour, images, fonts, file size

Wim Feringa

B.1 WEBSITE DESIGN

An Internet site is more than just a couple of pages with contents stitched together with some hyperlinks. The basis is the information that has to be transferred. A site designer's main concern is that the information will reach the user in the best way possible, which requires a logical structure of the site and an intuitive navigation without dead ends or wrong turns. Good graphic design should make it a pleasure to visit the site but the design has to serve the message, not the other way around. Another important element is the loading time. Web users are notoriously impatient. Especially if they are just surfing they are unlikely to wait more than 10 seconds for a page to load.

A well-structured site makes it easy to find the information. The starting point is to choose the best way to organise the information and the interface. The interface has to guide the visitor through the site, allowing full control over what to see. The structure and interface have to be carefully thought out before the actual design of the site can start. The importance of the graphic design of the site should not be underestimated. This can be compared with the packaging of goods in a shop. If a product is packed in a well-designed and carefully produced box, the customer assumes that the product is also of high quality. For websites it is not really different, users will assume the contents to be unreliable if the site is amateurish and poorly designed (Siegel, 1996). Site designers also need to bear technical aspects in mind, above all the need to make sites compact and fast loading.

B.1.1 Site architecture

A site has to be designed with goals, an intended audience and their expectations in mind. It makes a difference if the aim is to sell products, to educate, to provide information or just to entertain (see also Figure 8.2). It is a good idea to make as exhaustive a list as possible of the types of people who might be interested in the site and the reasons they may have for visiting it, e.g. "Students need statistical information about the country" or "Tourists – ask for maps of a certain region". In the same list try to indicate the expected hardware-software-connection speed for each group of visitors, with emphasis on those who are the main aim of the site. The last part is important for determination of the possible or required functionality the site needs to have.

After defining the type of visitors you want to serve mainly and the type of site it has to be, it is time to work out the contents. Come up with a list of all the subjects that need to be in the site such as home page, search, help, products, company information and so on. The contents should be organised in a hierarchy and relationships between the different subject groups and separate elements need to be made, since they need to be linked at a later stage.

A good rule of thumb is to avoid scrolling (or at least not more than 10%) on an opening page. Information is best split up to occupy not more than one page at a time. Links to related pages can be used to keep users active and interested, remembering that a website is normally not designed to be read sequentially, like a book. When there is sequential information, make sure the audience has no other choice than continuing to the next page. Certain long documents are not really suitable for presentation on the web, but are better downloaded or printed out, so that they can be read later.

The user has to be able to find his or her way around the site easily, so subject groups and their relations have to be organised in a logical structure. The easiest way to establish this is to make a model of the structure of the site with all pages and their internal and external links (Figure B.1). This site model is meant for the designer and shows all possible internal and external links.

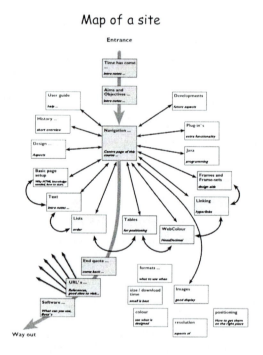

Figure B.1 A site structure model.

This model is different from site maps that we often see on websites as tools for navigation or orientation. A site map for the user shows the content of a site and the logical place where information can be found, but gives no further indication on content relationships. In the site map all different subjects are active links, leading to the correct page within the site (Figure B.2).

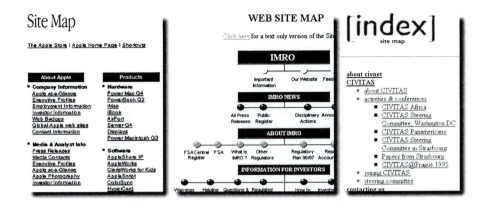

Figure B.2 Examples of three different types of site maps:
left, Apple (*URL B.1*); centre, IMRO (*URL B.2*); right, CIVITAS (*URL B.3*).

Another aspect of site design is the site scheme, which is not the same as the site model. The scheme shows how navigation through the site is organised, whereas the model shows contents and relations among the separate pages and links. Basically we can distinguish three types of site schemes: flat, balanced and deep *(URL B.4)*.

The flat scheme (Figure B.3) works with one main menu from where all (or most) pages can be reached *(URLs B.5 & B.6)*. Mostly there is a way back to the home page but it is sometimes forgotten, especially at sites, where not enough thought has been given to the future growth of the site. This kind of site grows organically until the whole site needs complete restructuring, requiring a lot of time. It is better to have a structure that is designed in such a way that there is space for growth without the need for a completely new design.

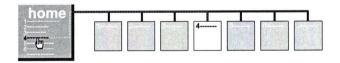

Figure B.3 A flat organisation scheme.

The flat scheme is not wrong per definition and it will work with simple sites. For more complex sites it is better to use sub-menus, located one step deeper in the scheme. This approach has the advantage that information can be better organised in subject groups. The visitors have a clear picture of where they are within the site and they can find their way easily to other places.

Once a website is started, more and more information will be added, with new subjects on new pages and new links. If this kind of growth has not been anticipated the organisation scheme might come to look like the one in figure B.4. At a certain moment the visitor will get lost in this deep organic growth, where information may not be logically grouped and many menus open new ones deeper and deeper into the scheme. Such a site will ultimately require a re-design (re-ordering and re-grouping of the information), perhaps even a split into new sites. This costs a lot of time, but may be inevitable since it is often very difficult to forecast in advance how a site will grow. The balanced scheme (Figure B.5) allows well for growth. This type groups related pages together but still allows easy access back to the home page, while it also allows the developer to expand the site easily.

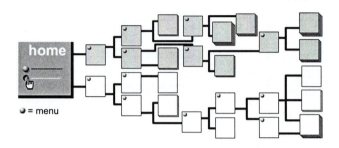

Figure B.4 A deep organisation scheme.

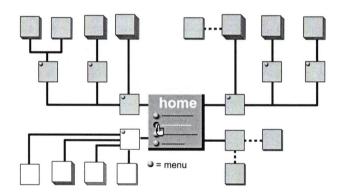

.**Figure B.5** A balanced organisation scheme.

Naturally, website designers wish the target groups to visit their site. Direct e-mailing takes time, advertising in newspapers and journals costs money. The best way is simply to make use of the many search engines on the Web (see also Chapter 3). Search engines need to know the location (the URL), some keywords for which the users can search and a short description of what the site is about. For the location of the site it is mostly sufficient to submit the address. The rest of the information, keywords and description, are located in the HTML files. This information could be different for each subject group or even for every single page. Therefore, the designer has to think carefully about the words that people might enter in a search engine to get the Internet pages with the required data, so those are exactly the words needed on the HTML page.

A website designer imposes a degree of uniformity on the site. For example it is wise to offer the information in the same format, meaning that the structure of the text is uniform throughout the site. This is also a good tool to give pages the same feel and a recognisable appearance. Dead-end pages should be avoided, so there should be a way to continue or at least to get back to the home page. Websites can be entered at any page, so the user needs to have an immediate idea of what the whole site is about. If it is desirable that users revisit the site regularly, which is usually the case for commercial sites, for example, the site has to be regularly updated.

B.1.2 User interface

Users interact with a site by moving and clicking the mouse and by entering information via the keyboard. These interactions are prompted by the graphic user interface of the site and by visible links within a page. The best interfaces do not give rise to doubts or mislead the user. A good interface is intuitive, so that every link, all icons and other elements give the response that one logically expects. Unfortunately, there is not really such a thing as a "fool-proof" interface, so designers have to presume an "average user". Professional website designers and others have studied how people react to and interact with an interface and many books have been written on the subject (for example, search for "gui" on the site of Amazon.com). Less experienced designers can learn a lot from the good practice of others, without directly copying of course.

There are many conventions used in software interfaces, such as what is to be found under the <file> and <edit> menus or what particular keyboard combinations do, such as <Ctrl> "x", <Ctrl> "c" and <Ctrl> "v". Other conventions much used on websites include a right-pointing arrow meaning "go on" and a left-pointing arrow meaning "go back". Buttons, icons and small graphic symbols can help to give the user a quick overview of what to find where. A well-designed button, icon or graphic not only says more than ten words, but also occupies less space on the page (Figure B.6).

Figure B.6 Well-designed buttons need no text.

Especially when these elements are used consistently throughout the whole site, the user can fully focus on the contents and will not be distracted by the interface. The interface should at be very transparent to the user. When this stage is reached the user will feel in full control. Not only consistency in type of elements but also in location of the elements on the page will make the navigation easy and logical. When the user knows where to find navigation tools and where they will lead to, he or she will find the desired information much more quickly. A very good aid for the visitors to a site is that they get an indication of where they are within the structure of the site. Often a system is used where at the top of the page the position of the page in relation to the home page is indicated, as in figure B.7.

Home / webdesign / interface / gui

Figure B.7 Where am I?

For navigation within a site there are basically two approaches, text-based (*URL B.3*) and graphic-based (*URL B.7*). Many sites combine the two (*URL B.8*). The advantage of a fully text-based navigation is the speed of downloading, but there is not much space for a nice design. Only if the users really need the information they are likely to stay, otherwise they will probably move on to another more attractive site. Furthermore, a well thought out and pleasant looking design will also give an impression of reliability. Good design is basically simple and direct. There should be no "noisy" distracting backgrounds to interfere with the readability of the text, especially to users with less than perfect vision. Also, many users are still restricted to a screen resolution of 480x640 pixels, so either the whole content has to fit on this screen or scroll bars have to open access to the parts that fall outside the screen. The only way to evaluate a site design is to send it over the Internet to volunteer testers, using various configurations and modems.

B.1.3 Page design

A web page is completely different from a page in a magazine or book. It consists of elements such as HTML files (perhaps organised in frames), contents from databases and web objects that trigger an event or activity. Web objects define areas that can receive mouse events, enabling JavaScript functions and hyperlinks. These dynamic elements transform a static HTML page into an active page with contents that can be derived from calculation, database search, or some other dynamic means (*URL B.9*). Despite these special dynamic aspects of a web page, many users print out interesting information they find on a site. This is partly because reading large amounts of text directly from a monitor can be tiring, partly because users can file the paper pages and write comments on them. For this reason it is best to include basic information on a web page such as the company name and address, the title of the text and the URL of the home page. All these elements will guide the user (or persons who receive a copy) at a later stage back to the source of the information.

For information on paper, whether it be textual or graphic, the dimensions are known and a design can be adapted to that but for screen display there are too many variables to be certain of the outcome. Dimensions of an Internet page can hardly be given, because too much depends on the configuration of the hardware at the client side. The design can be made in such a way that all information fits in one screen without scrolling, but the amount of information that can be put in one screen is so limited that contents often have to be distributed over more screens. The user will have to go from one page to another, which gives the user less overview of the total amount of information on the subject. It is also known, however, that Internet users hate scrolling and most only look at what they see first and do not go further. Their curiosity needs to be tickled to encourage them to scroll on. Scrolling down appears to be reasonably acceptable but scrolling across is not advisable and a combination of the two should be avoided, certainly for text-based pages. For maps on the Internet scrolling is not always avoidable, but it is best to try to fit the map to the minimum screen size defined at the start of the site design. The horizontal width is especially important, defined in pixels, not in centimetres or inches, since the screen dimensions are also based on pixels. Grids are often used to assist in page design in order to keep the pages of a site graphically consistent.

As for interface design, it is advisable to keep the page design simple and to avoid eye-catching backgrounds and too many colours and fonts. The information on the page must be clearly legible and the user should not be distracted. The website of the Royal Thai Survey Department (*URL B.10*) is an example of an overly complex, "noisy" page design. Research on how pages are read and what attracts one's attention gives some useful guidelines for web page design (e.g. Garcia, 2000 and Gavier, 2000). Some general hints have been derived from this research, based on the fact that readers:

- enter at the upper-right, continuing to top left, bottom left and leaving at bottom right and consequently do not pay too much attention to the centre;
- first look at headlines, photos and illustrations, especially large ones with high contrast;
- like subheads (easier to scan for information);
- do not like large blocks of text;
- do not like to scroll (therefore the most important information should be placed at the top of the page);
- like an illustration and its description next to each other;
- can be "pulled" into the site by using many ways of entering the same item (links, bullets, buttons, etc.).

B.2 INTERNET COLOUR

Most website designers prefer to make use of colour. The question really is whether colour on the Internet is still a special issue, given that computers are now powerful and modern video cards can definitely show thousands of colours. However, the designer must design for the target group and some target groups may still be equipped with a "minimum" configuration, i.e. with a display limited to 256 colours. The difference between a 256 colour (8 bit) and a 16.8 million colour (24 bits) display is shown in Figure B.8. Of course these black and white reproductions of the two screen dumps do not show the effect on colours, but they do give an indication of what happens to smoothly graduated tints. The full-colour effect can be seen on the website related to this book (*URL B.11*).

The monitor is the interface from the computer to the user and is the only possible way to actually see information on the World Wide Web. A monitor screen basically consists of a very fine pattern of red (R), green (G) and blue (B) phosphor dots. To produce multicoloured images, three electron guns are needed, one for each of RGB. The dots glow when electrons hit them. Different combinations of RGB give different colour impressions. For example, when only the red dots are glowing we see only red on the display, when red and green are glowing, we experience this as yellow. When all three are glowing, we see white and when no dot is hit by a gun the display remains black. All other combinations give other colours. The system described above is the one used in the standard Cathode Ray Tube (CRT). Various other types of display under the general name of Flat Panel Displays (FPDs) have been and are being developed, for example for laptop computers, but they are all based on the RGB combination.

Figure B.8 The difference between 8 bits and 24 bits display.

If each dot can be either "on" or "off" the result is 8 possible colours. To achieve more colours the strength of the electron stream must vary. The number of possible intensity variations depends on the video card in the computer. A 24-bit display reserves 8 bits (one byte) for each of RGB. Per colour there are 2^8=256 intensity variations, leading to $256^3 = 16\ 777\ 216$ different colours on the display. The problem with this system is that a coloured image requires a very large amount of data to represent it (24 bits for every pixel). This system is commonly called "true colour".

A somewhat different approach is to make a colour list or index. A common system allocates one byte per pixel to list the colours, allowing the use of 256 different colours in one image. Nowadays it is common to find two bytes allocated for the colours, allowing 65 536 colours. In fact, many website designers now assume that users will have this capability. However, in order to keep file sizes small, it is still best to limit the colours to 256.

A problem occurs when matching colours on a web page with the colours available in the user's own system, if set to 256 colours. Some colours may not be available on the user's system, which responds by "dithering", i.e. mixing pixels of the colours that do exist in order to achieve the best possible match with the intended colour. The amount of dithering required and the dither patterns themselves vary among the different systems. The result is that the dithered colours no longer appear smooth. Put another way, the web page designer does not have full control of how the user will see the chosen colours.

To tackle this problem, Netscape developed the "Web Safe Colour Palette", that has become an industry-wide standard. This palette is based on six intensity levels of RGB, leading to $6^3 = 216$ colours. On a percentage scale these intensity levels are 0, 20, 40, 60, 80 and 100, while on a scale of $0 - 255$ the corresponding values are 0, 51, 102, 153, 204 and 255. Any colour defined in this palette will appear true and smooth on any system using any browser. The remaining 40 colours (256 minus 216) are used by the operating system itself. The 216 Web Safe colours can be modelled in the form of a cube (Figure B.9).

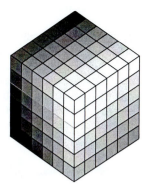

Figure B.9 One view of the Netscape Colour Cube,

In a HTML file, colours are not specified in percentages or on a scale from 0 – 255. Instead the so-called hexadecimal system is used. This has the advantage that any number from 0 to 256 can be expressed as a two-digit number. In this system the numbers 0 to 15 are expressed by the numbers 0 to 9, followed by A, B, C, D, E and F. The decimal number 16 is then expressed as the hexadecimal number 10. The decimal numbers 0, 51, 102, 153, 204 and 255 become (as hexadecimal numbers) 00, 33, 66, 99, CC, FF. In HTML an example of a Web Safe colour is "#99FF33", which is equivalent to 153R, 255G, 051B (on the scale 0 - 255) or, as percentages, 60%R, 100%G, 20%B. Maps and other images made using "flat" Web Safe colours will retain these when saved in the GIF format but may not appear smooth (see next section and Appendix A).

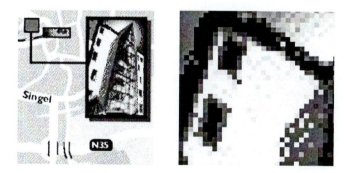

Figure B.10 A dithered image and a detail.

In a photographic image many colours are used to build up the image, more than the possible 256 of the indexed system or the 216 of the Web Safe palette. When a photograph has to be shown in less than the amount of colours needed to display it correctly, two things can happen: the image will be "dithered" (Figure B.10) or "banding" (Figure B.8) will take place. For dithering the operating system simulates the missing colours (not available in the palette) by creating random patterns of pixels of available colours. The viewer will experience the combination as an integrated colour. However, this dithering makes the file size much bigger (Appendix A, GIF file format).

The technique known as "antialiasing" can generate intermediate, "blended" colours. A screen image is built up of tiny pixels (picture elements). The result is that any line or sharp edge that is not perfectly horizontal or vertical will appear jagged when looked at closely. This jagged edge can be removed by adding pixels along it, intermediate in colour between the object and the background (Figure B.11). This antialiasing technique is very commonly used for text. The disadvantage of antialiasing is that the extra colours created make the file size bigger.

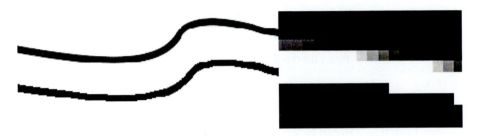

Figure B.11 The difference between an antialiased (smooth) and an aliased (hard) line.

B.3 IMAGE FORMATS AND FILE SIZE

Internet data have to be transferred over telephone lines. A modem is used for communication and for display there is a computer with a monitor. All these parts of the digital path have their restrictions in operation speed. A chain is only as strong as its weakest link, which is also the case here. In any event, the website designer should ensure that the site itself is not a weak link. Essentially, the designer tries to produce a high quality site within the smallest possible file size.

As was mentioned earlier, interesting sites contain not only text but various kinds of images, perhaps including maps. The graphic data formats for images result in large files, so for transport over the Internet the files are normally compressed. GIF and JPEG are the two most used formats for images and they are both fully supported by the browsers in common use. For descriptions of both formats see Appendix A. The compression technique used for the GIF format makes it most suited for images that have flat colours (many maps fit into this category). GIF compression makes use of indexed colours and this means that the designer has good control over the final number of colours and the final file size. When using the antialiasing technique the number of graduated colours selected has a great influence on file size, so it is best to choose the lowest number which still gives an acceptable result. It is anyway always best to test a GIF compression with varying parameters until a result is achieved which is slightly better than the minimum acceptable, and then to check it by actually sending the file over the Internet.

When using JPEG the designer controls the quality rather than the colour palette itself. This compression technique is very well suited for photographs or maps with many graduated tints (see Appendix A). As for GIF, the minimum acceptable quality is best found by trial and error. When a high compression rate is desired for rather large images with many (blending) colours, JPEG will give better quality and smaller files than GIF. The format is not really suitable for line images and it is less efficient than GIF for images containing only a few colours. For these reasons most maps are best compressed using GIF. Examples of different compression rates can be seen on the website that accompanies this book *(URL B.11)*.

Small files will download quickly but this is no guarantee for fast display. As soon as GIF or JPEG images arrive at their destination, they are decompressed. For display on the monitor they are unpacked and recalculated to the system's own settings. So when the setting is 24 bits, a coloured image of, say, 300x200 pixels and maybe only 5Kb as a compressed file will become 180Kb finally for the system to display. This becomes serious for large background images, for example. If a background has to fill the whole screen, let us say 1024x768 pixels, the final file size for display will be approximately 2.4Mb, even when the GIF or JPEG file measures only 35Kb. Not every PC owner is fortunate to have a video card with an endless amount of RAM installed on it to give a fast display of such a large amount of data. For these reasons, a user may set his system to 256 colours (with all the limitations involved) when using the Internet.

B.4 FONTS

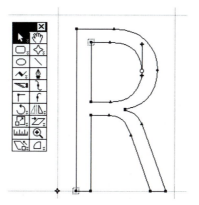

Figure B.12 Vector description of a character.

<font face="Times New Roman" size="3" color="#003300"> will be a familiar code for those who have ever made a web page, needed to identify type, size and colour of the font to be used in the web page. When a specific font is identified in a web page, the font description installed on the user's system will be used for displaying the page. The font "Times New Roman" is a standard font installed on every operating system and therefore never generates display problems. The designer can be sure that his or her carefully developed page will be displayed using the intended font. If a rarely used font is specified, not available on the user's system, then it will be replaced with the default font of the browser, very often in fact Times New Roman. This replacement may wreck a carefully designed page.

There are basically two types of fonts used at the moment: PostScript fonts and TrueType fonts. Both are defined as vectors, following the principle illustrated in Figure B.12. All characters that make up a font are installed in one file in a font directory as part of the system and every font installed at the user's side can be "called" by the designer of a page for use in the final display. The difficulty for designers is that they do not know what fonts are installed on the user's PC. He can only hope that the defined font will be there. The designer can include in the code some alternative, similar fonts, e.g. <font face="Helvetica, Arial">, to attempt to keep the design looking much the same.

Figure B.13 A decorative font and a symbol font.

As soon as decorative and symbol fonts are used (Figure B.13) designers can be almost certain that most users will not have these fonts installed. Then, they have two options, either to generate a GIF image of each of the desired words or symbols and place them on the web page, or to "embed" the font in the web page, i.e. to send the font description file over to the user. If there is a lot of text or symbols the first option tends to generate large files. Also, if the type or symbol images are enlarged the pixel structure becomes visible. The embedded font has the advantage that even a very large type remains sharp and does not cost extra memory, since the same vector data is used to generate types of all sizes.

Font embedding is made possible at this moment by a company called "Bitstream" that has produced TrueDoc *(URL B.12)*. This is more a method of font transport via the Internet than a new type of font. With the application "WebFont Maker" a TrueType font or a PostScript font is converted into an encoded description based on the outline description. A licence is needed to use or distribute most fonts. The font description files are decoded at the user's end by a Java application, residing on the server of the host. Microsoft, together with Adobe, has produced a new format called OpenType, which is not only a new format but also a way of delivering fonts over the Internet, comparable with TrueDoc. More information on this subject can be found on the websites of Microsoft *(URL B.13)* and Adobe *(URL B.14)*. Microsoft also has an application available named "Web Embedding Fonts Tool" (WEFT), to produce embedded fonts from unprotected (free) fonts. Note that these embedded fonts can only be viewed with the Internet Explorer browser.

Figure B.14 Antialiased and aliased text.

OpenType is not yet part of the standard Internet language and cannot yet form an integral part of maps sent over the Internet. Maps, together with their symbols and text, have to be rasterised first. When rasterising symbols or type the pixel dimension is always an important issue, compared to the size of the text or symbol. If the symbol or text is too small it will become unrecognisable after rasterisation. Figures B.14 and B.15 show that text needs to be at least about 10 pixels high (capital height) for it to remain recognisable any smaller causes possible confusion (e.g. the letter "a" might look like "3"). The minimum text size should be approx. 7pt. When creating a web map on a Macintosh platform select a minimum of 9pt. This will show up on a PC platform as approximately 7 pt. This is due to the fact that Mac fonts are physically smaller than their PC versions. Note that if you are rasterising text and symbols to be inserted at different sizes in a page or on a map it is best to rasterise the largest rather than the smallest size.

Figure B.15 Different sizes, antialiased and aliased.

There is not really a rule of thumb for minimum sizes nor for readability in combination with backgrounds; this has to be judged on screen for every single situation. When the symbol or text is located on a white or light background, the size can be smaller than against a "noisy" or dark background. Bear in mind that contrast is an important element influencing the readability of symbols or text. The choice of font is also very important. Not every font is suited for screen display and not every font is good for all sizes. The two basic font types are serif and sans-serif with for each many variations such as plain, bold, italic, bold italic, capital, lower case, condensed, expanded and in many colours and shades. Sans-serif styles such as Arial are perhaps less readable than serif styles such as Book Antiqua for blocks of text, but they are good for names on maps since they remain legible at small sizes.

Arial	Arial	Book Antiqua	Book Antiqua
AvantGarde	AvantGarde	Century Schoolbook	Century Schoolbook
Frutiger	Frutiger	Garamond	Garamond
Futura	Futura	Goudy	Goudy
Helvetica	Helvetica	Times	Times

Figure B.16 Sans-serif and serif fonts in a pixel size of 10.

Even within the serif and sans-serif styles, some fonts are more readable than others, as illustrated in Figure B.16. The most readable fonts at small sizes have an "open" character with a relatively large x-height. Note that the antialiasing technique can fill up small open spaces as in the letters "a" and "e". For this reason very small text is often best left aliased it will also produce a smaller GIF file. On web maps capital lettering may improve the readability of individual letters, as with an identical point size, they are larger and more open than lower case letters. This is in contradictory to paper maps, on which lower case lettering is preferred. Capital letters used in blocks of text may reduce reading speed, but probably it does not make much difference for single words such as names on a map.

URLs

Site Architecture
URL B.1   The Apple site map <http://www.apple.com/find/>
URL B.2   The IMRO site map <http://www.imro.co.uk/guide.htm>
URL B.3   The CIVITAS site map <http://civnet.org/sitemap.htm>
URL B.4   The Yale Web Style Guide, general reference
           <http://info.med.yale.edu/caim/manual/sites/site_design.html>
URL B.5   A flat structured website (Ethiopian Mapping Authority)
           <http://www.telecom.net.et/~ema/ema.htm>
URL B.6   Another flat structured site (Argentine Military Geographic Institute)
           <http://www.igm.gov.ar/>

(Graphic) User Interface
URL B.7   Graphic based navigation on the OEEPE site <http://www.oeepe.org/>
URL B.8   The Survey Department of Brunei uses a combination of graphics based
           and text based navigation <http://www.survey.gov.bn/>

Page Design
URL B.9   Web Objects information page <http://entropy6.stt.msu.edu/cgi-bin/
           WebObjectsExamples/WebScript/DynamicElements>
URL B.10   The Royal Thai Survey Department <http://schq.mi.th/rtsd/>

Colour and images
URL B.11   8 and 24 bits display <http:/kartoweb.itc.nl/webcartography/webbook/>

Fonts
URL B.12   System independent fonts <http://www.bitstream.com/webfont.index.html>
URL B.13   Web embedded fonts <http://www.microsoft.com/typography/default.asp>

URL B.14 Open Type technology <http://www.adobe.com/support/salesdocs/9c3a.htm>

REFERENCES

Garcia, M., 2000, Design myths <http://www.poynter.org/vjold/DTC/mario.html> (accesed 12.02.2000)
Govier, W., 2000, The eyes have it. Visual hierarchy helps readers <http://www.abobe.se/web/columns/govier/981022.wg.html> (accessed 05.07.2000)
Siegel, D., 1996, *Creating Killer Websites*, (Indianapolis: Hayden Books).

INDEX

DATE DUE